Research and Perspectives in Neurosciences

For further volumes:
http://www.springer.com/series/2357

Stanislas Dehaene • Yves Christen
Editors

Characterizing Consciousness: From Cognition to the Clinic?

Springer

Editors
Stanislas Dehaene, PhD
Professor at Collège de France
INSERM-CEA
Cognitive Neuroimaging Unit
91191 Gif sur Yvette
France
stanislas.dehaene@cea.fr

Yves Christen, PhD
Fondation IPSEN pour la
Recherche Therapeutique
65 quai George Gorse
92650 Boulogne Billancourt Cedex
France
yves.christen@ipsen.com

ISSN 0945-6082
ISBN 978-3-642-18014-9 e-ISBN 978-3-642-18015-6
DOI 10.1007/978-3-642-18015-6
Springer Heidelberg Dordrecht London New York

Library of Congress Control Number: 2011929691

Cover design: WMXDesign GmbH, Heidelberg, Germany

Printed on acid-free paper

Springer is part of Springer Science+Business Media (www.springer.com)

Introduction: Recent Advances in Consciousness Research

Consciousness was long considered as the "holy grail" of cognitive psychology and neuroscience: a vague and uncertain goal, so remote as to seem almost entirely out of reach. Today, however, the perspective has changed. Characterizing the computational architecture and neurobiological mechanisms underlying consciousness remains a major unsolved problem in cognitive neuroscience, but it has become an area of intense research. Thanks to new advances in stimulation paradigms, brain imaging techniques, and neuronal theorizing, the issue now appears to be empirically addressable. Yet a major challenge still confronts these novel empirical and theoretical proposals: will they be able to help clinicians confronted with patients in coma or vegetative state? Can they help define novel diagnostic or even therapeutic tools?

In the present book, which is the outcome of a Fondation Ipsen meeting held in Paris on May 3rd 2010, 13 renowned neuroscientists and clinicians examine whether consciousness research is ripe for applications, from cognition to the clinic. The diversity of empirical research is impressive, and the theoretical convergence is quickly growing. At the cognitive level, paradigms such as backward masking, binocular rivalry or change blindness, together with quantitative probing of the subject's introspective state of mind, are helping understand the extent of subliminal processing and the point where conscious processing starts. Brain imaging techniques, combined with novel analysis tools such as the new method known as multi-voxel pattern analysis, provides a window into the underlying brain state. A coordinated state of synchronized activity, emerging relatively late after the stimulus was presented and involving cardinal nodes in 'associative' cortical areas including prefrontal and parietal cortices, is frequently associated with conscious-level perception. Similar long-distance networks emerge spontaneously in the awake resting state, although whether they are necessarily associated with conscious experience remains debated. Many of these results appear compatible with the theory of a global neuronal workspace, which proposes that a distributed set of neurons with long-distance axons are involved in the global information broadcasting underlying reportability and what is experienced as a conscious state.

Most novel perhaps is the possibility to study consciousness in non-human primates. The ability of global networks to enhance their communication through phase synchronization is increasingly understood at the electrophysiological level.

Furthermore, new paradigms now ask whether animals possess meta-cognitive abilities, such as a self-monitoring of their competence in a task, and electrophysiologists now examine the underlying neuronal networks.

Turning to clinical applications, brain imaging in the resting state or in carefully crafted stimulation paradigms holds the potential to address three questions of central importance: Is a given patient conscious? Will he ever recover consciousness? And what will be his cognitive state if he does? Brain stimulation paradigms, whether cortical or in deep-brain nuclei, can alter the state of consciousness and may improve communication in some ‘minimally conscious’ patients.

In summary, consciousness research appears to be on the verge of concrete clinical applications. We hope that the present book will serve as an up-to-date survey of this exciting field.

Gif sur Yvette, France Stanislas Dehaene
Boulogne-Billancourt, France Yves Christen

Acknowledgments

The editors wish to thank Sonia Le Cornec for the organization of the meeting and Mary Lynn Gage for the editing of the book.

Contents

Contributors

Moran Cerf Computation and Neural Systems, California Institute of Technology, Pasadena, CA 91125, USA; Department of Neurosurgery, University of California Los Angeles, Los Angeles, CA 90095, USA; School of Business, New York University, New York, NY 10012, USA, moran@morancerf.com

Jean-Pierre Changeux Collège de France, 11 Place Marcelin Berthelot, 75005 Paris, France; Institut Pasteur CNRS URA 2182, Institut Pasteur, 75015 Paris, France, changeux@pasteur.fr

Camille Chatelle Coma Science Group, Cyclotron Research Centre, University of Liege, Liège, Belgium

Antonio Damasio Brain and Creativity Institute, University of Southern California, Hedco Neuroscience, Building Room 126, 3641 Watt Way, Los Angeles, CA 90089, USA, damasio@college.usc.edu

Stanislas Dehaene INSERM, Cognitive Neuroimaging Unit, Gif sur Yvette 91191, France; CEA, DSV, I2BM, Neurospin center, Gif sur Yvette 91191, France; University Paris 11, Orsay, France; Collège de France, 11 Place Marcelin Berthelot, 75005 Paris, France, Stanislas.Dehaene@cea.fr

Michel Desmurget Centre de Neuroscience Cognitive, UMR 5229, CNRS, Bron, France, Université Claude Bernard, Lyon 1, Lyon, France

Esteban A. Fridman Centre de Neuroscience Cognitive, UMR 5229, CNRS, Bron, France, Université Claude Bernard, Lyon 1, Lyon, France

Pascal Fries Donders Institute for Brain, Cognition and Behaviour, Radboud University Nijmegen, Nijmegen, The Netherlands; Ernst Strüngmann Institute (ESI) in Cooperation with Max Planck Society, pascal.fries@esi-frankfurt.de

John-Dylan Haynes Bernstein Center for Computational Neuroscience Berlin, Charité – Universitätsmedizin Berlin, Berlin, Germany; Max-Planck-Institute for Human Cognitive and Brain Sciences, Leipzig, Germany, haynes@bccn-berlin.de

Roozbeh Kiani Department of Neurobiology, Stanford University, Palo Alto, CA, USA

Steven Laureys Coma Science Group, Cyclotron Research Centre, University of Liege, Liège, Belgium Department of Neurology, Centre Hospitalier Universitaire Sart Tilman, Liège, Belgium

Michael Mackay School of Clinical Medicine, University of Cambridge, Cambridge CB2 0SP, UK

Lionel Naccache Neurology Department, Hôpital de la Pitié-Salpêtrière, Assistance Publique/Hôpitaux de Paris, Paris, France; CRICM, UMRS 975 Neurospychology and Neuroimaging Unit, Hôpital de la Pitié-Salpêtrière, Paris, France, lionel.naccache@psl.aphp.fr

Adrian M. Owen Centre for Brain and Mind, Room 224, Natural Sciences Building, The University of Western Ontario, London, Ontario, N6A 5B7, Canada, adrian.owen@uwo.ca

Mathias Pessiglione Institut du Cerveau et de la Moelle épinière, INSERM-CNRS, Université Pierre et Marie Curie, Paris, France, mathias.pessiglione@gmail.com

Marcus E. Raichle Departments of Radiology, Washington University School of Medicine, Neurology, Neurobiology and Biomedical Engineering, 4525 Scott Avenue, Room 2116, St Louis, MO 63110, USA, marc@npg.wustl.edu

Caroline Schnakers Coma Science Group, Cyclotron Research Centre, University of Liege, Liège, Belgium, C.Schnakers@ulg.ac.be

Michael N. Shadlen HHMI and Department of Physiology and Biophysics, University of Washington, Seattle, WA 98115, USA, shadlen@u.washington.edu

Angela Sirigu Centre de Neuroscience Cognitive, UMR 5229, CNRS, Bron, France; Université Claude Bernard, Lyon 1, Lyon, France, sirigu@isc.cnrs.fr

Herbert S. Terrace Department of Psychology and the New York State Psychiatric Institute, Columbia University, New York, NY, USA, terrace@columbia.edu

Thilo Womelsdorf Department of Physiology and Pharmacology, University of Western Ontario, London, Ontario, Canada; Donders Institute for Brain, Cognition and Behaviour, Radboud University Nijmegen, Nijmegen, The Netherlands, thiwom@imaging.robarts.ca

Missing Links in the Evolution of Language

H.S. Terrace

Abstract The evolution of language is an intractable problem if it is assumed (a la Chomsky) that language emerged full-blown at some point during the ~ 6,000,000 year period that elapsed between the divergence of the hominid line from chimpanzees, our closest living ancestor. It is intractable because it fails to recognize the role played by the evolution of conversation, and its antecedent, non-verbal conversation, a uniquely human phenomenon. For that we have to go back to the time when bipedalism replaced knuckle walking. Bipedalism resulted in a reduction in the size of the pelvis and the consequent reduction in the size of the birth canal, a birth canal that could not accommodate an infant with a brain larger than ~1,000 cc. As a result, human infants were born with a brain and skeletal system that were much less developed than those of non-human primates. To survive, the human infant required support from her mother for a period of time that was considerably longer than that needed by the offspring of other non-human primates. As a result, the human infant spent lots of face-to-face contact time with her mother, sharing eye gaze, another uniquely human activity. I hypothesize that those interactions allowed the infant to perceive and consciously compare the contents of her and her mother's mind. From there it is a relatively small step to joint attention, yet another uniquely human activity, which provided a basis for the assignment of names to events and objects.

It's not the same thing to talk of bulls, as it is to be in the bullring

Spanish proverb

All animals communicate, but only humans communicate with language. Animal communication is limited to a small number of inflexible, involuntary and predictable signals about basic needs and emotions, e.g., signals that communicate anger,

H.S. Terrace (✉)
Department of Psychology and the New York State Psychiatric Institute, Columbia University, New York, NY, USA
e-mail: terrace@columbia.edu

S. Dehaene and Y. Christen (eds.), *Characterizing Consciousness: From Cognition to the Clinic?* Research and Perspectives in Neurosciences,
DOI 10.1007/978-3-642-18015-6_1,

a foraging site, the appearance of a predator, fear, a desire to mate, etc. (Searcy and Nowicki 2005). Those signals cannot be modified by substituting and existing one for another or by inventing a new one. By contrast, human languages enable its practitioners to communicate voluntarily about any imaginable topic, e.g., politics, the fourth dimension, plans for the weekend, childhood memories, a recent novel, beauty, truth, the latest gossip, etc. (Jackendoff 2002; Pinker 2007).

The origin of language, the most recently evolved and the most complex form of communication, remains a mystery. Here, I will argue that there are two major obstacles to a viable theory of the evolution of language. One assumes a discontinuity between language and other forms of communication; the other does not. Those who argue for a discontinuity have noted that the main function of language is thinking and that thinking is primarily a manifestation of our superior problem-solving intelligence. Adherents of this view also claim that language differs so radically from other forms of communication that its basic features could not have evolved by natural selection (Chomsky 1965).

Like Darwin, those who hold the continuous point of view argue that, when the gap between two species seems too large, one must find the numberless gradations that intervene. But since language leaves no fossils, many students of language concluded that the problem was insoluble because it would prove impossible to find those gradations (Gould and Lewontin 1979; Pinker 1994). Glaringly absent from either position is the importance of particular social skills and how they co-evolved with language. Most linguists agree that the main function of language is social. Until recently, however, the social functions of language have been largely ignored. As I will maintain throughout this chapter, social skills that evolved with language are a necessary condition for the evolution of language.

The social skills in question evolved as a consequence of decidedly unsocial events. They were an indirect product of the evolution of bipedalism, which occurred approximately 6 million years ago (mya). Bipedalism resulted in a smaller pelvis and a concomitant reduction in the size of the birth canal. Approximately 1.5 mya, the increased volume of the brain created a mismatch between brain size and the size of the birth canal. This impasse was resolved by limiting the growth of the human brain in the womb to insure that the head could pass through the birth canal. However, once the new infant emerged, she was more helpless than any other non-human primate at birth. To survive, human infants required approximately 10 months of daily care from their mothers, the longest such period of any non-human primate. I will argue that the intense social interaction that took place during that 10-month period was the foundation for the evolution of language.

1 Theories of the Evolution of Language

Before elaborating that view, it would be helpful to review earlier attempts to formulate theories of the evolution of language. In response to the publication of *The Origin of Species* (Darwin 1859), various members of the European intelligentsia

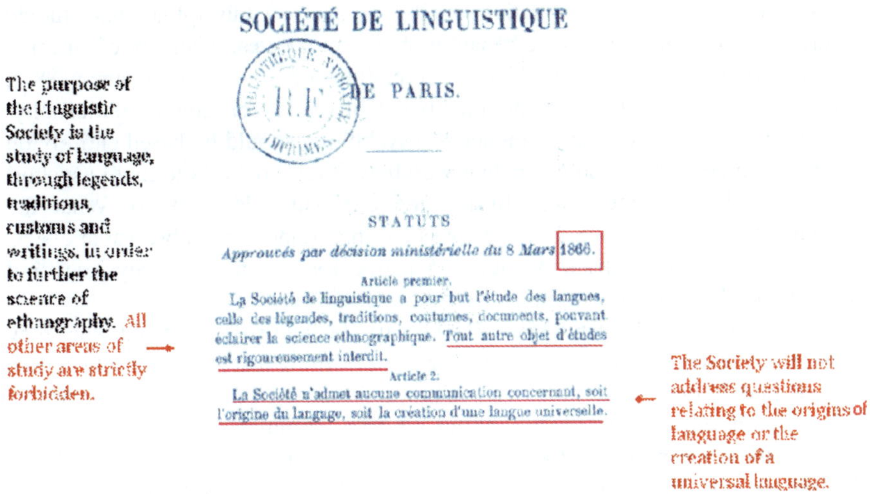

Fig. 1 Edict issued by the Linguistic Society of Paris banning further discussion of the evolution of language

proposed what have come to be known as onomatopoetic theories of the evolution of language. These theories, which were given appropriate names, like the bow-wow, the pooh-pooh, and the yo-he-ho (Kenneally 2007) theories, assumed that the first words a human uttered were imitations of natural sounds; for example, words like, *moo*, *bark*, *hiss*, *meow*, and *quack-quack* referred, respectively, to cows, dogs, snakes, cats and ducks. The obvious and fatal problem for all onomatopoetic theories was that, even if true, they could only account for a miniscule portion of the words of a basic vocabulary. They could not, for example, account for such common words as *same*, *beauty*, *truth*, *male*, *fish*, *tree*, *flower*, etc.

In response to the onslaught of these and other uncritical theories of the evolution of language, the Linguistic Society of Paris issued an edict in 1866 that banned all further discussions of that topic. An English translation of the Society's ban appears in Fig. 1. With minor exceptions, it was very effective. Indeed, more than a hundred years elapsed before psychologists, biologists, linguists, philosophers, neuroscientists and anthropologists showed an interest in that topic. What changed? As it turned out, quite a bit. The major developments were behaviorism, its demise, the cognitive revolution and ape language projects.

2 Behaviorism

In his *Principles of Psychology*, Herbert Spencer argued that, just as the theory of evolution was used to explain anatomical changes in different species over time, it applied with equal force to changes in animal behavior (Spencer 1886). That idea fostered an interest in comparative psychology, a new discipline that studied differences and similarities in the behavior of various species. After a brief flirtation with anthropomorphic and cognitive interpretations of behavior (Romanes 1884), most comparative psychologists adopted behaviorism as their guiding philosophy.

Behaviorists believed that a science of psychology should be based entirely on observable independent and dependent variables. Their goal was to mold psychology into a hard science like physics and chemistry (Hull 1943). What the behaviorists failed to notice, however, was the importance of unobservable events in the very sciences they tried to emulate. Imagine, for example, how physics would fare without the concepts of gravity or atomic particles.

During the first half of the twentieth century, behaviorists practiced what they preached by restricting themselves to theories of learning that relied entirely on two similar models of conditioning. The first was based on Pavlov's discovery that a dog would salivate in response to an arbitrary stimulus, such as a bell, that signaled the appearance of food [classical conditioning (S→S) (Pavlov 1927)]. The second was based on experiments in which Skinner trained rats to make an arbitrary response, such as pressing a bar, to obtain food [instrumental conditioning (R→S) (Skinner 1935)]. A common feature of virtually all conditioning theories of that era was the assumption that all learned behavior, in both animals and humans, including language, could be reduced to S→S or R→S associations.

Consider, for example, how a behaviorist would explain the way you find your hotel in a strange town. By trial and error, you learn which way to turn at seven choice points: left at the post office, right at the bank, right at the gas station, left after crossing a bridge, left at the playground, right at the library, and right after the high school. Your memory of which way to turn at each physically distinct choice point is all you need to return to your hotel. In that sense, learning a sequence of turns requires nothing more than the skills that a rat needs to navigate a seven-choice point maze.

As shown in Fig. 2a, responding correctly at the first choice point ensures that you will next encounter the second choice point, and only that choice point; responding correctly at the second choice point, ensures that you will next encounter the third choice point, and only that choice point, and so on, until you get to your hotel. What you learn can be characterized as a sequence of S–R associations, $\mathbf{S_1{:}R_1 \rightarrow S_2{:}R_2 \rightarrow S_3{:}R_3 \rightarrow S_4{:}R_4 \rightarrow S_5{:}R_5 \rightarrow S_6{:}R_6 \rightarrow S_7{:}R_7 \rightarrow S^R}$ (Skinner 1938), where $\mathbf{S_n}$ is a symbol for a particular choice point and $\mathbf{R_n}$ is a symbol for the correct response at that choice point. It is important to note that, with the exceptions of $\mathbf{S_1}$ and $\mathbf{S^R}$, each choice point functions as a reward for the prior response and as a discriminative stimulus for the next response.

Remarkably, Skinner's theory of grammatical sentences, which appeared in *Verbal Behavior* (1957), followed the same logic as that used by a rat to learn a maze. The first word of a sentence provided a cue for the second, which in turn provided a cue for the third, etc. Although, in hindsight, Skinner's theory of sentence construction seems simplistic, it should be viewed in the context of an extraordinary string of successful experiments that demonstrated the ease with which the principles of instrumental conditioning could be applied to just about

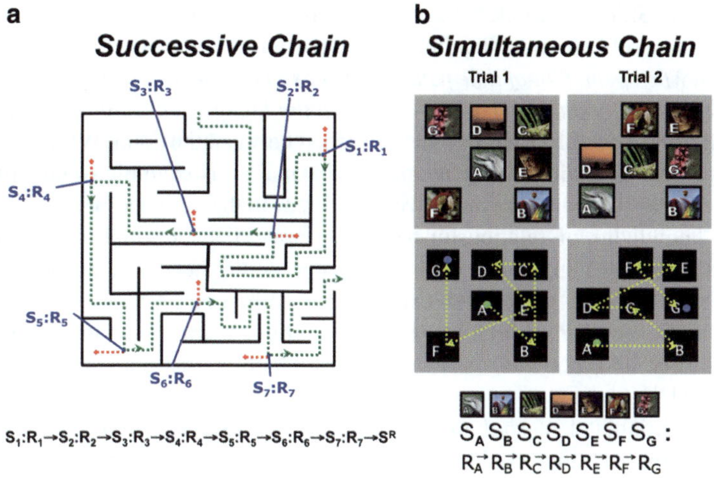

Fig. 2 (**a**) Basic plan of a traditional maze. Each choice point is separated, both spatially and temporally, from the others. A run through the maze is characterized as a *successive chain* that is composed of a series of S–R associations. (**b**) Simultaneous chaining paradigm. Each panel depicts a trial during training on a simultaneous chain. The configuration of the items, typically photographs, varies randomly from trial to trial. The photographs at the *bottom* of the figure show, for the benefit of the reader, the correct order in which a subject must respond to obtain a reward. The *bottom portion* of each panel depicts the route that a subject must follow (*yellow arrows*) with respect to the items shown in the *top panel* to receive a reward. Barring a 1 in 5,040 guess, a simultaneous chain must be learned by trial and error. A reward is presented only after the subject has responded to each item in the correct order. An error at any point of the sequence ends the trial. See text for additional details

any behavior in such disparate areas as behavior therapy (Masters and Rimm 1987), instruction by teaching machines (Skinner 1959), testing the effectiveness of particular drugs (Spiegel 2003) and teaching pigeons to guide missiles (Belzung and Le Pape 1994; Spiegel 2003), to name but a few. The success of this approach, in particular Skinner's imaginative applications of the principles of instrumental conditioning (Skinner 1953), fostered a climate in which it seemed possible to teach an animal to master any skill, including language. It was therefore not surprising that many of Skinner's students began programs whose goals were to obtain empirical evidence to support his theory of language.

3 Chomsky's Critique of Verbal Behavior and Behaviorism

Chomsky, arguably the most important linguist of the twentieth century, was not impressed by Skinner's achievements using instrumental conditioning. In a highly influential review of *Verbal Behavior* (Chomsky 1959), Chomsky exposed some fatal flaws in Skinner's model of sentence construction.[1] For example, a 5-year-old child has no problem understanding the sentence – *A father told his son that Ted Williams, who spent a lot of time fishing with custom made fishing rods and who was also a famous pilot during World War II, had the highest seasonal batting average of all* – despite a gap of 22 words that intervened between the noun, Williams, and its predicate, had. Chaining theory is unable to explain this example of an embedded sentence because it can only move from one item to the next and because it cannot explain knowledge of relationships between non-adjacent words. Chomsky also provided examples in which chaining theory could not even account for the meaning of adjacent words, for example, in ambiguous sentences and phrases such as *they are visiting firemen* or *the shooting of hunters*, or in a nonsensical sentence such as *green ideas sleep furiously*, which is grammatically correct, but nonsensical.

In his critique of *Verbal Behavior*, Chomsky explored other grammars, of greater complexity that could account for the meaning of sentences that were beyond the grasp of chaining theory and that could also account for the effortless manner in which speakers of all languages transform simple declarative sentences into other tenses; for example, changing *John throws the ball to Bill*, to its negative, interrogative, passive and passive negative forms, respectively: *John is not throwing the ball to Bill, Does John throw the ball to Bill? John had thrown the ball to Bill*, and *John hadn't thrown the ball to Bill*. We will not pursue more recent models of grammar that Chomsky and his students have proposed and, indeed are still proposing, because they are applicable to grammars that appeared long after language evolved. More relevant are Chomsky's arguments that grammar is the distinguishing feature of language and that an innate Language Acquisition Device (LAD) does the heavy lifting in the evolution of language.

The LAD is an example of what cognitive psychologists refer to as a module; a functional portion of the brain that is dedicated to the control of a particular cognitive ability, e.g., perception, language, memory, etc. (Fodor 1983; Cosmides and Tooby 1992). The LAD has many functions that helped to compensate for what Chomsky referred to as the impoverished stimulus. The problem is that the verbal stimuli that a child experiences seem too meager to explain a child's linguistic knowledge. Most words are learned after a single exposure to its referent. Indeed, a mere 2 years after a child utters her first word, her vocabulary increases dramatically to thousands of words, reaching a rate that has been estimated to be as high as one new word/hour. More impressive is a child's mastery of grammatical rules without any formal instruction. For Chomsky, these facts can only be explained by the LAD, an innate module of the brain that contains the abstract rules of a universal grammar.

---

[1]Lashley (1951) made a similar argument and also suggested a hierarchical, as opposed to a linear, structure of learned responses in which particular nodes controlled the spacing of particular subordinate responses. These and related ideas have been described in detail by others and will not be elaborated in this chapter (Anderson and Bower 1974). [Gardner]

Chomsky and other linguists have asked, quite reasonably, how could so major a mechanism as an LAD have evolved by natural selection during the approximately 6 million years that have elapsed since humans and chimpanzees split from a common ancestor? To be sure, the point of Chomsky's question is not whether language evolved but how natural selection could account for the sudden appearance (as measured in evolutionary time) of the most complex form of natural communication known to man (Chomsky 1986). Anything less than a single mutation wouldn't do because, just as a quarter or a half a wing would not be of any help to flying birds that evolved from flightless birds, a quarter or a half of a grammar would not be of any use to our ancestors when attempting to communicate with language. If there were such stages, Chomsky asks, what kind of grammars would they be?

Given the complexity of the LAD, Chomsky argued that it could not have evolved in what was the equivalent of an eye blink in evolutionary time. Instead, he argued that the LAD took over a part of the brain that was used by some other function, such as navigation, by a process known as exaptation (Hauser et al. 2002). In this view, the LAD did not owe its existence to natural selection because it was already in place when the need to generate and comprehend grammatical utterances arose. A number of biologists, citing Darwin's concept of pre-adaptation (Williams 1966; Gould 1977; Gould and Vrba 1982), have argued that many structures first appeared as exaptations. Wings, for example, are considered to be exaptations of structures whose original functions were predation and/or thermoregulation. It is important to recognize, however, that exapted structures are just as susceptible to natural selection as any other structures. Primitive wings, for example, which could propel birds in flight for only very short distances, could be modified by natural selection for long-distance flying. Similarly, natural selection could have modified an exapted LAD to accommodate various environmental pressures. But that explanation begs the question: what were those pressures?

4 Ape Language Projects

One of the surprising developments during the last half of the twentieth century was the appearance of a large number of projects that questioned the view that language, and grammar in particular, was uniquely human. Why the sudden appearance of these projects? One obvious reason was Chomsky's controversial conclusion that only humans can use language to communicate. That view provoked students of animal behavior to find evidence that he was wrong. Indeed, Skinner's influence on psychology was at a peak at the time and, as mentioned earlier, Skinner's achievements created an expectation that any type of behavior could be trained. Another factor was the observation that the human and chimpanzee vocal apparati differed significantly (Lieberman 1968, 1975) and that a chimpanzee's inability to articulate human phones might explained various failures to teach home-reared infant chimpanzees to speak either English or Russian (Hayes 1951; Kellogg and Kellogg 1933; Ladyna-Kots 1935).

In each case, the idea was to reverse those failures by shifting from a vocal to a visual medium of communication. The species that participated in these studies varied widely: parrots (Pepperberg 1991), dolphins (Herman et al. 1984), orangutans (Miles 1983), gorillas (Patterson 1978) and chimpanzees, both *Pan troglodytes* (Gardner and Gardner 1969; Premack 1971; Rumbaugh et al. 1973; Terrace et al. 1979) and *Pan paniscus* (Savage-Rumbaugh 1994). Most of these studies presented evidence that non-human animals can communicate by using language. If true, such claims would obviously complicate theories of the evolution of language. A review of these claims, with which I disagree, will clarify why language remains a thorn in the side of the theory of evolution. In the interest of brevity, I will focus on experiments that studied chimpanzees, the most popular subject of these projects. My conclusions, however, apply equally to other species.

The Gardners sought to reverse earlier failures by teaching a chimpanzee a spoken language by using American Sign Language (ASL), a natural language used by thousands of deaf Americans. ASL was the main medium of communication between Washoe (an infant female chimpanzee) and her caretakers and between the caretakers themselves while in Washoe's presence. Premack, who started an independent project at roughly the same time that the Gardners began theirs, taught the principal subject of his study (a juvenile female chimpanzee named Sarah) to use an artificial visual language consisting of plastic chips of different colors and shapes. Rather than waiting for language to emerge spontaneously, as one might with a child, Premack devised specific training procedures for teaching Sarah various atomic components of language.

Researchers studying ape language accepted as a given the prevailing working assumption of psycholinguistics that human language makes use of two levels of structure: the word and the sentence. In contrast to the fixed character of various forms of animal communication (e.g., bird songs that assert the presence of food or a readiness to mate and bee dances that specify the location of a food source with respect to the hive), the meaning of a word is arbitrary. One must keep in mind, however, that even though apes can learn substantial vocabularies of arbitrary symbols, there is no a priori reason to regard such accomplishments as evidence of human linguistic competence. Dogs, rats, horses, and other animals can also learn to produce arbitrary words to obtain specific rewards.

A second level of structure, one that assumes the word (Chomsky 1965), is generally regarded as the essential feature of human language. Sentences characteristically express propositions through words and phrases, each bearing particular grammatical relations to one another (actor, agent, object, and so on). Psychologists, psycholinguists, and linguists are in general agreement that a human language requires knowledge of a grammar. How else can we account for a child's ability to create an indefinitely large number of meaningful sentences from a finite number of words?

In an early diary report, the Gardners noted that Washoe used her signs in two or more... in 29 different two-sign combinations and four different combinations of three signs (Brown 1970). That report prompted Brown to comment, It was rather as if a seismometer left on the moon had started to tap out 'S-O-S' (p. 211). Indeed, Brown, who was the most eminent psycholinguist of his time, compared Washoe's sequences of signs to the early sentences of a child and noted similarities in the structural meanings of Washoe and children's utterances (agent-action, agent-object, action-object, and so on).

Other projects reported similar combinations of two or more symbols. Sarah produced strings of plastic chips such as *Mary give Sarah apple* (Premack 1976). Lana, a juvenile female chimpanzee, was trained to use an artificial visual language of lexigrams. Each lexigram, which is a combination of a specific geometric configuration and a specific colored background, was presented on the keys of a computer console or on a large visual display. After learning to use individual lexigrams, Lana learned to produce sequences of lexigrams such as *Please machine give M & M* (Rumbaugh et al. 1973). Because variations of these sequences occurred on the first occasion in which it was appropriate to use other signs or lexigrams, it was concluded that such sequences were actual sentences.

In an early diary report, Roger Fouts, one of Washoe's main trainers, described what is arguably the most famous sequence of signs generated by a chimpanzee (Fouts 1975). Washoe reportedly signed *water bird* after Fouts, asked her – *what's that?* – in the presence of a swan. What made this observation even more remarkable was the fact that Washoe did not have signs for specific water birds such as swans and ducks. It therefore seemed to Fouts that Washoe invented a way of conveying what she saw by a combination of signs.

Before we accept Fouts' rich interpretation of Washoe's sequence, it is necessary to rule out four simpler interpretations of *water bird*: (1) Washoe may have been prompted by Fouts to sign *water bird*; (2) Washoe may have signed *bird water*, but Fouts may have recorded Washoe's utterance in the order in which English-speaking people combine adjectives and nouns; (3) Washoe may have signed *water* and *bird* as two separate utterances; and (4) Fouts had previously trained Washoe to make the signs *water* and *bird* for food reward before the swan appeared. Washoe may have therefore signed *water bird* for a food reward without any specific understanding of what those signs referred to.

To eliminate such killjoy interpretations of this and other sentences that chimps were purported to have produced, I decided to start my own project centered around an infant chimpanzee I named Nim Chimpsky (Terrace 1979). Nim was a young male chimpanzee that, like Washoe, had been reared by human surrogate parents in an environment in which ASL was the major medium of communication. The goal of Project Nim was to obtain non-anecdotal evidence of the grammatical structure of sequences of signs produced by a chimpanzee. His trainers accomplished that goal by dictating the signs they observed, as well as their contexts, into miniature tape recorders and then transcribing the tapes after their session with Nim ended.

Over the course of 4 years, Nim's teachers recorded more than 20,000 combinations of two or more signs. Superficially, many of Nim's combinations appeared to be generated by simple finite-state grammatical rules (e.g., more + x; transitive verb + me or Nim). Indeed, many of Nim's multi-sign utterances

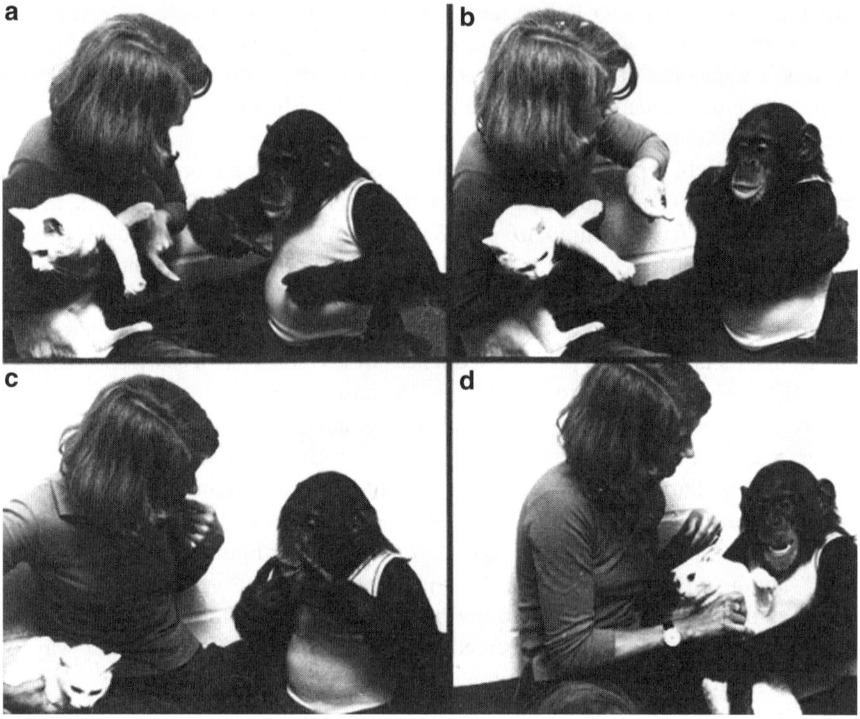

Fig. 3 Nim signing the linear combination *me hug cat* to his teacher (Susan Quinby). From Terrace et al. (1979b). Photo by H. S. Terrace

resembled a child's initial multiword utterances (Braine 1976; Bretherton and Bates 1979; Nelson 1981). All told, the corpus of Nim's utterances provided the strongest evidence to date of an ape's ability to create sentences.

As clear as that combinatorial evidence seemed, closer inspection showed otherwise. A frame-by-frame analysis of videotapes of Nim's signing revealed that his signs were nothing more than elaborate requests for rewards that he could not otherwise obtain and that there was no basis for interpreting the various sequences he produced as sentences (Terrace et al. 1979; Terrace 1979). The vast majority of Nim's signs, both in his single- and multi-sign utterances, occurred when his teachers withheld rewards until he signed. That insured that Nim would sign many multi-sign sequences. Nim's combinations were also full or partial imitations of signs that his teachers used as prompts. Virtually none of Nim's sequences were spontaneous. When, for example, Nim wanted to play with a cat his teacher was holding, he might sign *Nim cat, me cat, hug cat, Nim hug cat* before his teacher handed him the cat. Even then, videotape analyses showed that the teacher prompted Nim continuously as he was signing *who that?* or, *Nim want cat?* See, for example, Fig. 3, which shows what appears to be a three-item sequence, *me hug cat*. Each of those signs was triggered by a teacher's sign

Fig. 4 Nim being prompted to sign cat by his teacher (Susan Quinby). From Terrace et al. (1979b). Photo by H. S. Terrace

(lower left hand corner of the upper right and the two bottom panels). Figure 4 shows Nim's trainer drilling Nim on the sign *cat* before the cat was released from her box.

In hindsight, the meanings of Nim's signs were projections of his teachers, who directed all of their attention to his signing in the interest of obtaining objective records of each sign. So focused were they on Nim's signs that they were unaware of their nonspontaneous and imitative nature. It is, of course, true that young children imitate many of their parents' utterances. But, as shown in Fig. 5, the relative frequency of a child's imitated utterances is substantially lower than that of a chimpanzee. Further, although the imitative phase in children is transitory, Nim never moved beyond that phase (Terrace et al. 1979). As shown in Fig. 6, Nim's signing remained predominantly nonspontaneous and imitative, unlike the highly accelerated growth of a child's vocabulary at the end of Stage I of language acquisition (cf. Bloom et al. 1976; Brown 1973). Analyses of available films of other signing apes revealed similar patterns of nonspontaneous and imitative discourse [e.g., Washoe signing with the Gardners (Gardner and Gardner 1973) and her other teachers, and Koko signing with Patterson (Schroeder 1977)]. Miles (1983) performed a discourse analysis of videotapes of the orangutan Chantek's signing with his teachers and reported, there is no evidence ... that Chantek's multi-sign combinations ... are sentences (p. 53).

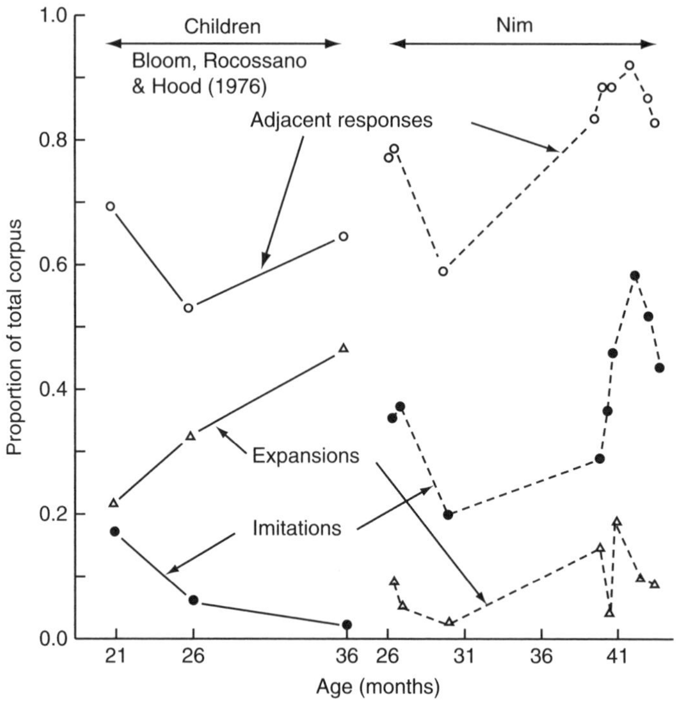

Fig. 5 Proportion of utterances emitted by children (*left-hand function*) and by Nim (*right-hand function*) that are adjacent to, imitative of or expansions of an adult's prior utterance. From Terrace et al. (1979b)

The conclusions of Project Nim were criticized on various methodological grounds by other investigators attempting to teach an ape to use sign language, e.g., Gardner (1981) and Patterson (1981). However, these investigators have not revealed enough of their own procedures to be sure that they were immune to the same criticisms (Terrace 1981, 1982). For example, while visiting the Washoe and Lana Projects, I noticed that the trainers on those projects regularly fed their chimpanzees with small pieces of food to motivate them to use the language on hand. Of greater interest is the fact that I could have reached the same conclusions by looking at videotapes of *other* chimps using ASL or, for that matter, as an armchair psychologist who had no experience trying to teach an ape to use language. All that would be needed to refute my conclusions would be a single, one-hour *unedited* video tape of a chimpanzee that was purportedly signing with a human companion in which *both* parties were visible in each frame. That tape would show the extent to which the trainers coached their chimps and whether those chimps were given tidbits of food to keep them motivated. During the 40 years that have elapsed since I published my conclusions, I have yet to see a single videotape that challenged them.

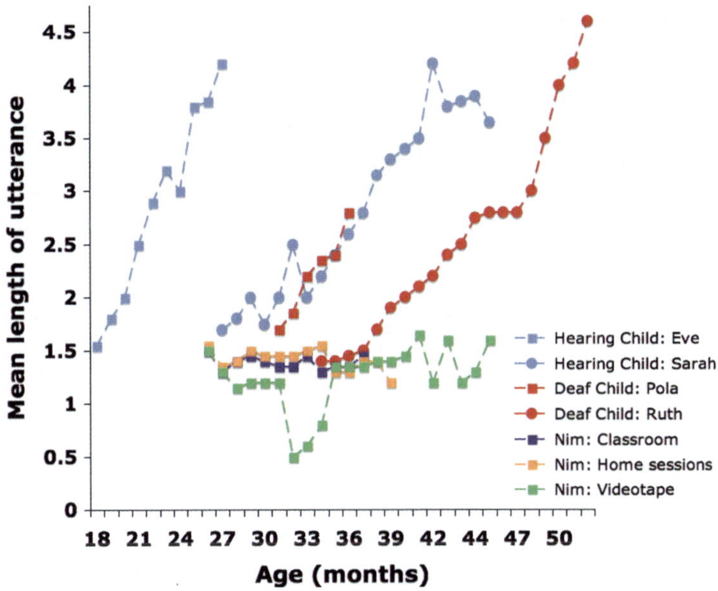

Fig. 6 Mean length of signed utterances of Nim and three deaf children and mean length of spoken utterances of two hearing children. The functions showing Nim's MLU between January 1976 and August 1977 (age 26–45 months) is based upon video transcript data. From Terrace et al. (1979b)

5 Rote Sequences Versus Sentences

Different considerations led to a rejection of the view that Sarah's and Lana's sequences were sentences. Thompson and Church (1980) analyzed a corpus of approximately 14,000 of Lana's combinations that were collected by a computer. They concluded that almost all of those combinations could be attributed to two non-grammatical processes: conditional discrimination and paired-associate learning. First, Lana learned paired associates, each consisting of a particular lexigram and a particular incentive (e.g., $lexigram_{apple} \rightarrow$ apple). Lana then learned conditional discrimination rules that determined which of six stock sequences she should produce. For example, if the incentive was in view of the machine that dispensed rewards, the stock sequence would take the form, *Please machine give piece of X, where X* refers to the symbolic member of the paired associate, e.g., the lexigrams *apple*, *music*, *banana*, and *chocolate*. If there were no incentive in view, the appropriate sequence would be, *Please put into machine X*. Typically, the paired-associate symbol was inserted into the last position of the stock sentence. Although Lana clearly understood the meanings of the paired-associates, in the sense that she could use them contrastively to make specific requests, there is no evidence that she understood the meanings of other lexigrams. Similar arguments have been made about the plastic symbols that Sarah used to produce her sequences (Terrace 1979).

5.1 Serial Learning in Monkeys

Further evidence of the non-sentential nature of Lana's (and Sarah's) sequences was provided by studies that showed that monkeys could learn arbitrary sequences of photographs, e.g., tree → bird → flower → person (Chen et al. 1991). The lists on which the monkeys were trained differed radically from traditional tasks that have been used to train animals to learn an arbitrary sequence (e.g., running through a maze). As described earlier, a subject must learn how to respond at successive choice points to navigate a maze. Because the choice points of a maze are experienced in isolation from one another, there is no need to represent one's position in the sequence while navigating the maze. It was sufficient to recognize each choice point as a particular stimulus of a *successive* chain and, to then respond appropriately.

Instead of training monkeys to learn successive chains, we trained them to learn *simultaneous* chains, a much more difficult serial task. In contrast to a successive chain, all of the choice points that comprise a simultaneous chain were displayed at the same time, typically on a touch-sensitive video monitor (Terrace 1984). Another important feature of a simultaneous chain was random changes in the configuration of the choice points from trial to trial to prevent subjects from learning a fixed series of motor responses by mapping each response to a particular location on the touch screen. Reward was withheld until the subjects responded to each item correctly. We then insured that subjects had to rely exclusively on *non*-differential secondary reinforcement (the positive or negative feedback that followed their responses to each of the first $n - 1$ items). Taken together, these features of a simultaneous chain forced a subject to maintain a representation of its position in the sequence before making each response. The following thought experiment, based on the seven-item simultaneous chains on which monkeys were trained, shows why.

Imagine trying to enter your seven-digit personal identification number (PIN) at a cash machine, say 9-2-1-5-8-4-7, on which the positions of the numbers were changed each time you tried to obtain cash. You could not enter your PIN by executing a sequence of distinctive motor movements, that is, by first pressing the button in the lower right corner of the number pad to enter 9, then the button in the upper middle position to enter 2, and so on. Instead, you would have to search for each number and keep track of your position in the sequence mentally as you pressed different buttons. As difficult as this task may seem, it would be far more difficult if you did not know your PIN and you had to discover it by trial and error. Any error ended that trial and resulted in a new trial on which the digits were displayed in a different configuration. Thus, to determine your PIN, you would have to recall the consequences of any of the 21 types of logical errors you could make while attempting to produce the required sequence. Further, you would have to determine the first six digits without getting as much as a penny from the cash machine. This is precisely the type of problem the monkeys had to solve at the start of training on each seven-item list of photographs. Instead of cash, they were given banana pellets.

As shown in Fig. 2b, the monkey's task was to respond to each item in a particular order, regardless of its spatial position. To learn each list, subjects had

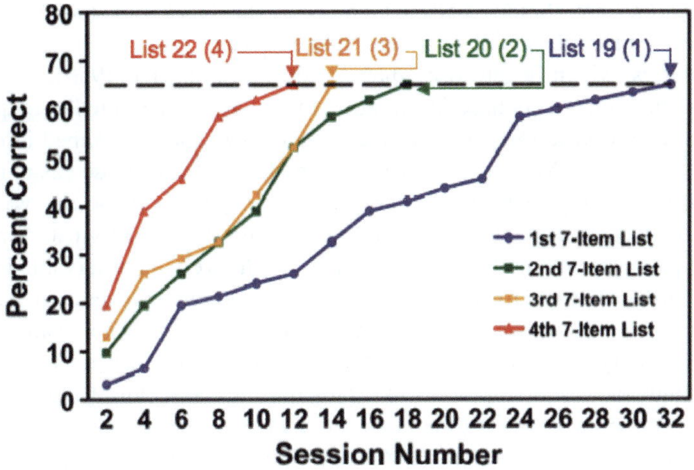

Fig. 7 Learning curves for seven-item lists. The mean accuracy of responding on each seven-item list during even-numbered sessions. The probability of executing a new seven-item list correctly by chance, assuming no backward errors, is 1/5,040 (1/7!). Note that the *abscissa* represents session (not list). From Terrace et al. (2003)

to make logical errors while determining the ordinal position of each item at the start of training on a new list. A logical error is the first incorrect guess a subject makes to a particular item at a given position of the list (e.g., responding to G on a seven-item list in the second position). By definition, each type of logical error can occur only once. Although logical errors are necessary for discovering the ordinal position of an item, repetitions of the same error are not. Logical errors are made to obtain information by virtue of their consequences. For example, G cannot be the second item because the trial was terminated. Repetitive errors occur because the subject has forgotten the consequences of an earlier logical error.

Subjects made progressively fewer errors during the course of learning new seven-item lists. The likelihood of guessing the correct order in which to respond to the items of a novel seven-item list (0.0005) was less than 1/5,000th. Each monkey nevertheless needed progressively fewer sessions to satisfy the high accuracy criterion on these lists (65% correctly completed trials in one session). As can be seen in Fig. 7, subjects needed on average 31.5, 17.5, 13, and 12.25 sessions to master seven-item lists 1, 2, 3, and 4, respectively (ranges: 21–55, 11–25, 11–19, and 7–17, respectively).

6 The Cognitive Revolution

Behaviorists believed that mental processes, if they did indeed exist, took place in a black box, which was not their concern. To address those processes, they passed the buck to physiological psychologists. The behaviorist's job was simply to understand the correlations between inputs to the black box (conditioned and unconditioned stimuli) and the behavior that emerged.

Computers, which began to be mass-produced during the 1950s, changed all that. Computer engineers used information theory not only to predict output from input but also to describe particular mental processes that intervened in concrete and objective terms. Mental processes could no longer be dismissed as figments of the imagination. Of greater importance, models of cognitive events that were based on the architecture of a computer, and the software it ran, had greater predictive power than behavioral models that were based exclusively on observable input and output (Gardner 1985). This was especially true of cognitive processes that required a memory of events that were not present at the time of recall. Without any means to describe memory, behaviorists had to rely on such unwieldy concepts as the action of a stimulus over time, a process that falls outside the realm of physics. To the chagrin of many behaviorists, the addition of memory between the input and the output of the black box allowed cognitive psychologists to increase its predictive value to a higher level than behaviorists could achieve. Behaviorists could not ignore that fact because the ability to predict is a hallmark of the hard sciences.

One unintended consequence of the cognitive revolution was the development of a parallel discipline, animal cognition. Although behaviorists would regard animal cognition as an oxymoron, it is ironic that most current investigators of animal cognition are former behaviorists. It would take us too far afield to review the impressive achievements of animal cognition, but it would be helpful to glance at some of those achievements, if for no other reason than to realize just how large a gap between animal communication and language remains even after these recently discovered abilities about the animal mind are taken into account. Examples of animal cognition can be found in experiments on concept formation (Wasserman and Bhatt 1992; Herrnstein et al. 1976), delayed, symbolic and successive matching-to-sample (Kendrick et al. 1981; Fagot et al. 2001; Wright et al. 1984), numerical ability (Brannon and Terrace 1998, 2000; Cantlon and Brannon 2006; Beran et al. 2008; Boysen 1993; Biro and Matsuzawa 1999; Hauser et al. 1996), serial learning (D'Amato and Colombo 1988; Merritt et al. 2007; Terrace et al. 2003; Swartz et al. 1991; Chen et al. 1997; Sands and Wright 1980), metacognition (Kornell et al. 2007; Shields et al. 1997), spatial learning (Spetch and Kelly 2006), relational learning (Cook et al. 1997; Young and Wasserman 1997), timing (Gibbon 1977; Meck and Church 1983; Breukelaar and Dalrymple-Alford 1998) and transitive inference (Treichler et al. 2003) among others [see Wasserman and Zentall (2006) for additional topics].

7 Missing Links

There is ample evidence that animals, and non-human primates in particular, are surprisingly intelligent when tested on a large range of problems. Monkeys, a species that has never been cited for its linguistic abilities, can learn rote sequences that are qualitatively similar to the sequences of signs, lexigrams or plastic chips that chimpanzees were trained to produce by rote. The failure of various projects to train chimps to learn language and the ability of monkeys to learn long sequences appear to be two sides of the same coin. The sole function of the sequences that both species acquired was to obtain various types of reward. In that sense, they are superficial examples of what linguists refer to as *proto-imperatives*, i.e., a verbal demand for a tangible reward (Bates et al. 1975).

They are superficial because proto-imperatives only have meaning in the context of *proto-declaratives*, i.e., utterances that assume a speaker and a listener and that occur because of the sheer joy that the speaker and the listener experience from exchanging ideas, recipes, gossip, news, weather, etc. Without proto-declaratives, there can be no exchange of information. As far as I am aware, there is no evidence that a non-human primate can produce or understand proto-declaratives or that communicating with declaratives can be taught. In brief, without proto-declaratives, there can be no language.

That conclusion should extinguish any hope of teaching language to a non-human primate. It does not, of course, imply that language did not evolve. As Darwin suggested, the key to explaining gaps that are too large for the theory of evolution to explain is to look for smaller gaps. In the case of intelligence, Darwin observed that:

> ... Everyone who admits the principle of evolution, must see that the mental powers of the higher animals...are the same in kind ... Thus *the interval between the mental powers of the higher apes and man ... is filled up by numberless gradations* (Darwin, The Descent of Man, 1871, p. 126) italics added

Fortunately, evidence from recent research on infant cognition and attachment theory provides new ideas for filling the gap between chimpanzees and humans with respect to language. The most important features of this evidence are (1) its reliance on *Homo erectus* and on some of our hominid ancestors as points of reference for the evolution of language, rather than the chimpanzee; and (2) its emphasis on social factors. Many linguists have recognized the importance of those factors but, given the difficulty of collecting such evidence, that recognition amounts to lip service.

7.1 Anatomical Changes in Hominids

The lack of fossil evidence has been frequently cited as a stumbling block for studying the evolution of language. Although there is no new evidence available that addresses this issue, it is possible to use existing evidence from anatomical changes that resulted from bipedalism and the steady increase in the size of the hominid brain.

An immediate consequence of bipedalism was a reduction in the size of the pelvis. A large pelvis was needed to support the four limbs of apes but, with only two limbs to support, the pelvis shrank and, with it, the size of the birth canal.

The main consequence was the limit on the size of an infant's head at birth. The approximate volume of the brains of our oldest known bipedal ancestor is 300 cc, as opposed to 1,300 cc for *Homo sapiens*. The size of the birth canal did not cause any problems until *Homo erectus* appeared, whose brain volume was approximately 900 cc, a volume that could just pass through the birth canal.

Nature's solution to this problem was to limit the size of a fetus at birth to insure the safe passage of the fetus' brain. However, giving birth to less developed infants meant that their brains were approximately 20% of the size they would reach as adults. By contrast, the brain of a newly born chimpanzee is 40–50% of its adult size. Since movements are controlled by the motor cortex, a small brain size also meant that human infants were less agile at birth than their chimpanzee counterparts.

The major consequence of the diminutive size of a human infant's brain is its greater need for maternal care. Since human infants are not able to crawl until they are approximately 8 months old, their mothers have to carry them for a much longer period than mothers of more developed primates. By itself, that increment in maternal investment would not necessarily have altered the cognitive development of a human infant. However, two other physical changes that also occurred during the era in which Homo erectus was the most prominent member of the homo species also contributed to the mix: one was the loss of bodily fur, the other, a whitening of the sclera of the human cornea.

The loss of bodily fur meant that infants could no longer cling to their mothers' hairs. Instead, they had to be held in their mothers' arms. During a 10-month period of relative helplessness, the human infant spent more time looking into her mother's eyes, which were made all the more salient by the contrast between a white sclera and a dark iris that was greater than in any other non-human primate. Given the long period during which human infants were held by their mothers, the many opportunities for mutual gaze created an important dynamic with uniquely human consequences.

In sum, there are four reasons to look to our homo heritage rather than our descent from apes to understand the evolution of language: upright posture, a small birth canal, loss of bodily fur and a white as opposed to a dark sclera. It should be clear that none of these differences was selected to enhance linguistic ability directly. In combination, however, they led to human infants that had special training during their infancy to form uniquely human social relationships with their peers, which led to language.

Although there is wide agreement about the uniqueness of a human infant's social skills, there is a lack of consensus about what they should be called. A Theory of Mind (TOM) is the most popular name but social scientists do not agree about what aspects of a child's cognitive development it should cover. As defined by Premack (Premack and Woodruff 1978) in a classic article about primate cognition, TOM referred to an agent's capability of representing another individual's mental states, so as to be able to understand, predict, and explain his behavior. Premack eventually concluded that there was not enough evidence to attribute a TOM to a chimpanzee but the term took root in studies of cognitive development in children. It was not long before questions like – Does a TOM require language? Can a TOM be attributed to newborn human infant or to animals? – arose. These have yet to be resolved, in part because some investigators have different requirements as to whether a complete suite of mental functions is needed before references to a TOM are legitimate. As a result, non-verbal and perhaps unconscious phenomena such as intersubjectivity and joint attention have been used synonymously with a full-blown TOM that requires consciousness and verbal ability. My own view is that, like intelligence, the concept of TOM has outlived its usefulness because it is too broad. What is important is to recognize that the significance of a TOM rests simply with the ability of one individual to recognize some aspect of another's mind. Once we have a basis for distinguishing different stages of its development, it is reasonable to assign specific names to a particular stage. To do otherwise would Balkanize a very important concept and would dilute its significance.

7.2 *Intersubjectivity*

The mental life of an infant during her first year has become a new frontier in developmental psychology. For example, we know that infants a few hours old can imitate facial and manual gestures, that 4-month-old infants can engage in spatial and numerical reasoning and that 9-month-old infants can direct the attention of other individuals to objects in her immediate environment. These achievements are all the more impressive because they show that infants can think without language.

Investigators of animal cognition could reply that there is nothing special about an infant's mental life because birds, rats, monkeys and chimpanzees can also think without language. There is, however, one aspect of our mental life of that appears to be exclusively human: the concept of self. The mere mention of self is likely to provoke controversy, if for no other reason than it lacks a simple definition. We nevertheless agree that a self exists, especially when contrasted with other. We can nevertheless also agree that, without the concepts of first and third person, it would be meaningless to discuss language. What we do not know, however, is the origin of the concept of self. One compelling hypothesis is that it develops from intimate exchanges between an infant and her mother soon after birth. Those who argue for the continuity of cognitive events claim that the concept of self is innate on the grounds that it must derive from some cognitive antecedent. In principle, I agree with the latter hypothesis but, given the state of the art, it is currently not feasible to pursue it. For our purposes, the origin of the concept of self does not matter because it seems clear that it is accentuated by the intimacy of the infant–mother relationship, and it is the outcome of that relationship that is most relevant to the evolution of language.

By convention, we will refer to the first year of an infant's life as intersubjective. Although the basis of that term may appear obvious, it is important to place it in a biological perspective. In 1929, John Bowlby, a pioneer in the study of the emotional development of human infants, documented similarities in attachment behaviors in different primate species and human cultures. In that respect, Bowlby viewed human attachment in the context of primate evolution as did Harlow and Hinde in later studies

The significance of Bowlby's insights were highlighted by Winnicut and Kline in object relations theory, a psychoanalytic theory that describes the process of the developing mind in relation to others in the environment, in particular with an infant's mother. Trevarthen (1977) and Stern (1977), both developmental psychologists, called attention to the importance of an infant's attachment to her mother.

Trevarthen and Stern independently emphasized the difference between an infant's ability to control inanimate objects and communication between two individuals. An infant can control and predict the outcome of moving an object, e.g., dropping it, by virtue of her innate knowledge of physics (Spelke et al. 1994). However, an infant has to rely on her experience to control and predict the outcome of communication between other individuals and herself. By the age of 3 months, there is much evidence that infants have begun to master each type of control. They look at, listen to and touch various inanimate objects but they smile, babble, cry, coo, laugh and point when they are in the presence of their mother. The mother's response is often predictable. She displays different facial expressions, engages in different displays of body language, talks in motherese, sings, etc. A dramatic example of such interchanges can be observed in a 4-month-old infant when she plays peek-a-boo with her mother.

Most important is the time the infant and the mother gaze at each other. The outcome of those interactions can produce different emotional reactions in the infant and the mother. Those reactions are essential for the regulation of an infant's emotional state. For example, an infant's expression of fear can be quickly replaced by the mother's smile and her expression of contentment can be quickly replaced by the mother's frown or by non-engagement. It is through these hour-by-hour interactions with her mother that the infant is said to develop a sense of other. She cannot always predict when her mother will smile or frown but she can recognize when her mother's emotional state differs from hers.

As far as I am aware, such interactions have not been observed in other primates. To be sure, there are examples of mutual eye gaze, laughter, etc., but there is no evidence that they can be sustained with anything that approaches the intensity of the mother–infant relationship in humans. Should we assume that the limited interactions of non-human infants and their mothers do not result in a concept of self and other? Of course not. But there is minimal evidence that the concept of self and other is not maintained as the infant grows into a juvenile and an adult. The most economical conclusion about the self–other distinction in non-human primates is that they rely on body language, rather than another's mental state, to predict a conspecific's behavior. Indeed, there is virtually no evidence that a non-human primate entertains the possibility that his conspecifics have minds.

7.3 Joint Attention

During the first 9 months of a human infant's life, her interactions with her mother are deictic, that is, they attend to each other rather than to objects in the environment. That phase is often referred to as *primary intersubjectivity* in contrast to the phase referred to as *secondary intersubjectivity*, which normally begins at the age of 12 months.

During secondary intersubjectivity, the infant and the mother begin to share attention to various objects in their environment, that is, they enter into triadic interactions with each other. Triadic interactions are often referred to by the more transparent term, joint attention. Discovered by Scaife and Bruner (1975), joint attention is regarded as the foundation of an infant's vocabulary. Since learning a vocabulary is the first step in learning a language, its significance should be obvious.

Joint attention requires eye gaze on the part of the infant and her mother to a particular object *and* the infant's acknowledgment that her mother is attending to that object. That acknowledgment has been interpreted as evidence that the infant feels that her perception matches her mother's. Joint attention is often preceded by shifts in the infant's eye gaze from the object to her mother and from her mother to the object. Those shifts in eye gaze have been interpreted as attempts by the infant to coax her mother to attend to the object in question.

Joint attention has rightly been regarded as the major cognitive achievement of infancy (Tomasello and Bellagamba 1999). Without joint attention, it is hard to imagine how vocabulary can be taught to an infant or, for that matter, how the infant can be taught to engage in any activity (Premack 1986). As Tomasello and Bellagamba (1999) and Hobson (2002) vividly describe, there could be no sharing of cognitive events; there would be no culture without joint attention.

The claim that joint attention is uniquely human has been questioned on the grounds that a similar phenomenon can be observed in non-human primates. To address those claims, it is important to distinguish between mutual eye gaze (Emery 2000) and what has erroneously been described as shared attention. In both instances, two primates have been observed to be gazing in the direction of the same object. What is not clear is whether either of those primates is aware of what the other primate sees. As emphasized by Povinelli, there is a fundamental difference between gazing and seeing (Povinelli and Vonk 2003). Gazing describes a behavioral event; seeing, a mental event. As in other instances of inferring a mental event, the evidence is a specific behavior that accompanies the event in question. In the case of joint attention, the infant's efforts to direct her mother's attention to a particular object by repetitively shifting her eye gaze between the object and her mother, and her satisfaction about achieving that outcome, are the basis of saying the infant sees the object in question. As far as I am aware, such behavior has not been observed in non-human primates.

8 Conclusions

1. Efforts to explain the evolution of language by bridging the gap between ourselves and chimpanzees, our closest living ancestors, failed for a variety of reasons.
2. Concerted efforts to train chimpanzees to learn non-vocal languages failed because the only function of the symbols they were taught to use was to obtain various types of reward.

3. Monkeys were able to learn long sequences of arbitrary items, an ability once thought to be important for creating sentences. Accordingly, sequence learning per se can be ruled out as a sufficient condition for producing sentences.
4. The shift from a quadrupedal to a bipedal posture, which occurred about 6–7 mya, resulted in a reduction in the size of the pelvis and a concomitant reduction in the size of the birth canal. The size of the birth canal placed a limit on the volume of an infant's brain that could pass through it.
5. When the brain reached that limit, most likely in *Homo erectus*, infants were delivered with brains that were less developed than those of other non-human primates. Accordingly, the main growth of the brain to its adult size occurred outside the mother's womb.
6. The fragile state of an infant with an underdeveloped brain necessitated intensive care on the part of the mother.
7. In *Homo erectus*, the loss of fur and a change in the color of the sclera of the cornea, from black or brown to white, also appear to have influenced how a mother reared her infant. Because the mother had no fur to which the infant could cling, the mother had to lift the infant to a higher position on her chest. In that position, the contrast between a dark iris and a white sclera increased the salience of the mother's eyes, making her eyes a more visible target for her infant's eye gaze.
8. The close face-to-face proximity of the infant and her mother facilitated their intersubjectivity, from which emerged the infant's concept of self and other.
9. At an age of approximately 12 months, a change occurred from a deictic relationship between the infant and her mother to a triadic relationship whereby the mother and the infant coordinated their attention to external objects.
10. The infant's sense that her perception of an object matched her mother's allowed them to refer to the object in the same manner. That led to the use of *words*, the first step in learning to use a language.

References

Anderson JR, Bower GH (1974) Human Associative Memory. Washington, D.C., Hemisphere Publishing

Bates E, Camaioni L, Volterra V (1975) The acquisition of performatives prior to speech. Merill-Palmer Q 21:205–226

Belzung C, Le Pape G (1994) Comparison of different behavioral test situations used in psychopharmacology for measurement of anxiety. Physiol Behav 56:623–628

Beran M, Evans T, Harris E (2008) Perception of food amount by chimpanzees based on the number, size, contour length and visibility of items. Animal Behavior 75:1793–1802

Biro D, Matsuzawa T (1999) Numerical ordering in a chimpanzee (Pan troglodytes): planning, executing and monitoring. J Comp Psychol 113:178–185

Bloom LM, Rocissano L, Hood L (1976) Adult–child discourse: developmental interaction between information processing and linguistic knowledge. Cogn Psychol 8:521–552

Boysen ST (1993) Counting in chimpanzees: nonhuman principles and emergent properties of number. In: Boysen ST, Capaldi EJ (eds) The development of numerical competence animal and human models. Erlbaum, Hillsdale, pp 39–59

Braine MDS (1976) Children's first word combinations. Monogr Soc Res Child Dev 41:1–96

Brannon E, Terrace HS (1998) Ordering of the numerosities 1–9 by monkeys. Science 282: 746–749
Brannon E, Terrace HS (2000) Representation of the numerosities 1–9 by rhesus monkeys. J Exp Psychol Anim Behav Process 25:31–49
Bretherton I, Bates E (1979) The emergence of intentional communication. New Dir Child Dev 4:81–100
Breukelaar JW, Dalrymple-Alford JC (1998) Timing ability and numerical competence in rats. J Exp Psychol Anim Behav Process 24:84–97
Brown R (1970) The first sentences of child and chimpanzee. In: Brown R (ed) Psycholinguistics. Free Press, New York, pp 210–211
Brown R (1973) A first language: the early stage. Harvard University Press, Cambridge
Cantlon J, Brannon E (2006) Shared system for ordering small and large numbers in monkeys and humans. Psychol Sci 17:401–406
Chen S, Swartz KB, Terrace HS (1991) Preliminary evidence for the development of learning set for 4-item lists by rhesus monkeys. Paper presented at annual meeting of the Eastern Psychological Association
Chen S, Swartz KB, Terrace HS (1997) Knowledge of the ordinal position of list items Tein rhesus monkeys. Psychol Sci 8:80–86
Chomsky N (1959) A review of Skinner's Verbal Behavior. Language 35:26–58
Chomsky N (1965) Aspects of the theory of syntax. MIT, Cambridge
Chomsky N (1986) Knowledge of language. Praeger, Westport
Cook RG, Katz JS, Cavoto BR (1997) Pigeon same–different concept learning with multiple stimulus classes. J Exp Psychol Anim Behav Process 23:417–433
Cosmides L, Tooby J (1992) Cognitive adaptations for social exchange. In: Barkow C, Tooby J (eds) The adapted mind: evolutionary psychology and the generation of culture. Oxford University Press, New York
D'Amato MR, Colombo M (1988) Representation of serial order in monkeys (Cebus apella). J Exp Psychol Anim Behav Process 14:131–139
Darwin C (1859) On the origin of species. Mercury, London
Darwin C (1871) The Descent of Man and Selection in Relation to Sex. Princeton, NJ, Princeton University Press
Emery NJ (2000) The eyes have it: the neuroethology, function, and evolution of social gaze. Neurosci Behav Rev 24:581–604
Fagot J, Wasserman EA, Young ME (2001) Discriminating the relation between relations: the role of entropy in abstract conceptualization by baboons (Papio papio) and humans (Homo sapiens). J Exp Psychol Anim Behav Process 27:316–328
Fodor J (1983) Modularity of mind. MIT, Cambridge
Fouts RS (1975) Capacity for language in great apes. In: Tuttle R (ed) Socioecology and psychology of primates. Mouton, Chicago, pp 371–390
Gardner BT (1981) Project Nim: Who taught whom? Contemporary Psychology 26:425–426
Gardner H (1985) The mind's new science: a history of the cognitive revolution. Basic, New York
Gardner BT, Gardner RA (1969) Teaching sign language to a chimpanzee. Science 162:664–672
Gardner BT, Gardner RA (1973) Teaching sign language to the chimpanzee: Washoe. The Psychological Cinema Register, University Park
Gibbon J (1977) Scalar expectancy theory and Weber's law in animal timing. Psychol Rev 84 (3):279–325
Gould SJ (1977) Ontogeny and phylogeny. Harvard University Press, Cambridge
Gould SJ, Lewontin RC (1979) The spandrels of St. Marco and the Panglossian program: a critique of the adaptationist program. Proc Roy Soc London 205:281–288
Gould SJ, Vrba ES (1982) Exaptation: a missing term in the science of form. Paleobiology 8:4–15
Hauser MD, MacNeilage P, Ware M (1996) Numerical representations in primates. Proc Natl Acad Sci USA 93:1514–1517
Hauser MD, Chomsky N, Fitch WT (2002) The faculty of language: what is it, who has it, and how did it evolve? Science 298:1569–1579

Hayes C (1951) The ape in our house. Harper, New York
Herman LM, Richards DG, Wolz JP (1984) Comprehension of sentences by bottlenosed dolphins. Cognition 16:129–219
Herrnstein RJ, Loveland DH, Cable C (1976) Natural concepts in pigeons. J Exp Psychol Anim Behav Process 2:285–302
Hobson P (2002) The cradle of thought: exploring the origins of thinking. Macmillan, London
Hull CL (1943) Principles of behavior. Appleton, New York
Jackendoff R (2002) Foundations of language: brain, meaning, grammar, evolution. Oxford University Press, Oxford
Kellogg LA, Kellogg WN (1933) The ape and the child: a study of environmental influence and its behavior. McGraw-Hill, New York
Kendrick DF, Rilling M, Stonebraker TB (1981) Stimulus control of delayed matching in pigeons: directed forgetting. Exp Anal Behav 36:241–251
Kenneally C (2007) The first word: the search for the origins for language. Viking, New York
Kornell N, Son LK, Terrace HS (2007) Transfer of metacognitive skills and hint seeking in monkeys. Psychol Sci 18:4–71
Ladyna-Kots N (1935) Infant chimpanzee and human child. Museum Darwinianum, Moscow
Lashley KS (1951) The problem of serial order in behavior. In: Jeffries LA (ed) Cerebral mechanisms in behavior. Wiley, New York, pp 112–136
Lieberman P (1968) Primate vocalizations and human linguistic ability. J Acoust Soc Am 44: 1574–1584
Lieberman P (1975) On the origins of language: an introduction to the evolution of human speech. Macmillan, New York
Masters JC, Rimm DC (1987) Behavior therapy: techniques and empirical findings. Harcourt, San Diego
Meck WH, Church RM (1983) A mode control model of counting and timing processes. J Exp Psychol Anim Behav Process 9:320–334
Merritt D, MacLean EL, Jaffe S, Brannon EM (2007) A comparative analysis of serial-ordering in ring-tailed lemurs (Lemur catta). Comp Psychol 121:363–371
Miles HL (1983) Apes and language: the search for communicative competence. In: de Luse J, Wilder HT (eds) Language in primates: perspectives and implications. Springer, New York, pp 43–61
Nelson K (1981) Individual differences in language development: implications for development and language. Dev Psychol 17:170–187
Patterson FG (1978) The gestures of a gorilla: language acquisition by another pongid. Brain Lang 12:72–97
Patterson FG (1981) Ape Language. Science 211:86–87
Pavlov IP (1927) Conditioned reflexes. Oxford University Press, London
Pepperberg I (1991) Language acquisition and form in a bilingual environment: a framework for studying birdsong in zones of sympatry. Ethology 89:1–28
Pinker S (1994) The language instinct. Morrow, New York
Pinker S (2007) The language instinct: how the mind creates language. Harper Perennial Modern Classics, New York
Povinelli DJ, Vonk J (2003) Chimpanzee minds: suspiciously human? Trends Cogn Sci 7:157–160
Premack D (1971) Language in a chimpanzee? Science 172:808–822
Premack D (1976) Intelligence in Ape and Man. Hillsdale, NJ, Lawrence Erlbaum Associates
Premack D (1986) Pedagogy and aesthetics as sources of culture. In: Gazzaniga MS (ed) Handbook of cognitive neuroscience. Plenum, New York, pp 15–35
Premack D, Woodruff G (1978) Does the chimpanzee have a theory of mind? Behav Brain Sci 1:515–526
Romanes G (1884) Mental evolution in animals. AMS, New York
Rumbaugh DM, Gill TV, von Glasersfeld EC (1973) Reading and sentence completion by a chimpanzee. Science 182:731–733
Sands SF, Wright AA (1980) Primate memory: retention of serial list items by a rhesus monkey. Science 209:938–940

Savage-Rumbaugh ES (1994) Kanzi: the ape at the brink of the human mind. Wiley, New York
Scaife M, Bruner J (1975) The capacity for joint visual attention in the infant. Nature 253:265–266
Schroeder B (1977) Koko: the talking gorilla. Home Vision
Searcy WA, Nowicki S (2005) The evolution of animal communication: reliability and deception in signaling systems. Princeton University Press, Princeton
Shields WE, Smith JD, Washburn DA (1997) Uncertain responses by humans and rhesus monkeys (Macaca mulatta) in a psychophysical same–different task. J Exp Psychol Gen 126:147–164
Skinner BF (1935) Two types of conditioned reflexes and a pseudo-type. J Genet Psychol 12:66–67
Skinner BF (1938) The behavior of organisms. Appleton, New York
Skinner BF (1953) Science and human behavior. Macmillan, New York
Skinner BF (1957) Verbal behavior. Appleton, New York
Skinner BF (1959) Teaching machines. Cumulative record. Appleton, New York, pp 158–182
Spelke ES, Katz G, Purcell SE, Ehrlich SM, Breinlinger K (1994) Early knowledge of object motion: continuity and inertia. Cognition 51:131–176
Spencer H (1886) The principles of psychology. Appleton, New York
Spetch ML, Kelly DM (2006) Comparative spatial cognition: processes in landmark and surface-based place finding. In: Wasserman E, Zentall T (eds) Comparative cognition: experimental explorations of animal intelligence. Oxford University Press, New York, pp 210–228
Spiegel R (2003) Psychopharmacology: an introduction. Wiley, Chichester
Stern DN (1977) The first relationship: infant and mother. Harvard University Press, Cambridge
Swartz KB, Chen S, Terrace HS (1991) Serial learning by rhesus monkeys. I: Acquisition and retention of multiple four-item lists. J Exp Psychol Anim Behav Process 17:396–410
Terrace HS (1979) Nim. Knopf, New York
Terrace HS (1981) Reply to Bindra and Patterson. Science 211:87–88
Terrace HS (1982) Why Koko can't talk. Sciences 22:8–10
Terrace HS (1984) Simultaneous chaining: the problem it poses for traditional chaining theory. In: Commons ML, Herrnstein RJ, Wagner AR (eds) Quantitative analyses of behavior: discrimination processes. Ballinger, Cambridge, pp 115–138
Terrace HS, Petitto LA, Sanders RJ, Bever TG (1979) Can an ape create a sentence? Science 206:891–902
Terrace HS, Son L, Brannon E (2003) Serial expertise of rhesus macaques. Psychol Sci 14:66–73
Thompson CR, Church RM (1980) An explanation of the language of a chimpanzee. Science 208:313–314
Tomasello M, Bellagamba F (1999) Re-enacting intended acts: comparing 12- and 18-month-olds. Infant Behav Dev 22(2):277–282
Treichler FR, Raghanti M, Van Tilburg DN (2003) Linking of serially ordered lists by macaque monkeys (Macaca mulatta): list position influences. J Exp Psychol Anim Behav Process 29:211–221
Trevarthen C (1977) Descriptive analysis of infant communicative behavior. In: Schaffer HR (ed) Studies in mother–infant interaction. Academic, London, pp 227–270
Wasserman EA, Bhatt A (1992) Conceptualization of natural and artificial stimuli by pigeons. In: Honig WK, Fetterman JG (eds) Cognitive aspects of stimulus control. Erlbaum, Hillsdale, pp 203–223
Wasserman EA, Zentall TR (2006) Comparative cognition: experimental explorations of animal intelligence. Oxford University Press, New York
Williams GC (1966) Adaptation and natural selection. Princeton University Press, Princeton
Wright AA, Santiago HC, Sands SF (1984) Pigeon and monkey serial probe recognition: acquisition, strategies, and serial position effects. Erlbaum, Hillsdale
Young ME, Wasserman EA (1997) Entropy detection by pigeons: Response to mixed visual displays after same-different discrimination training. J Exp Psychol 23(2):157–170

Consciousness as a Decision to Engage

Michael N. Shadlen and Roozbeh Kiani

Abstract Consciousness encompasses a variety of functions and properties, such as awakening, awareness, and subjective aspects of both perception and volition (e.g., qualia and authorship, respectively). It remains to be seen whether these diverse functions are related to one another through common neural mechanisms, and if so how. Here, we advance the thesis that the neural mechanisms that give rise to conscious states share features with neural mechanisms that underlie simpler forms of decisions. The neurobiology of decision-making provides detailed insight into how the brain deliberates and reasons from evidence to make choices. The underlying mechanisms, mainly studied in animals, could support a variety of complex cognitive functions that probably operate independently of many aspects of consciousness. For example, many complex decisions in humans rely upon wakefulness but not upon awareness or authorship. In animal studies, decisions are typically embodied: they can be described as selection among possible actions. By substituting "circuits" for "actions" in the preceding phrase, we generalize from "deciding to do" to "deciding to consider" or, more generally, "deciding to decide to...." This is an appealing notion from the perspective of brain evolution, because it allows us to recognize ideation as an elaboration of a simpler sensory-motor design. We propose that many of the functions of consciousness are simply ways of engaging the environment. Thus consciousness might be mediated by (non-conscious) decisions to engage, as in awakening, or to engage in a certain way, as when attaching narrative to action. Although the neural mechanisms underlying "decisions to engage" are unknown, they are likely to involve intralaminar (and matrix) thalamus and processes that "decide" to turn other circuits on. This idea invites an analogy between the functions of brain regions that project to

M.N. Shadlen (✉)
HHMI and Department of Physiology and Biophysics, University of Washington, Seattle, WA 98115, USA
e-mail: shadlen@u.washington.edu

S. Dehaene and Y. Christen (eds.), *Characterizing Consciousness: From Cognition to the Clinic?* Research and Perspectives in Neurosciences,
DOI 10.1007/978-3-642-18015-6_2, © Springer-Verlag Berlin Heidelberg 2011

matrix thalamus, including the "default system", and the role of parietal cortex in perceptual decisions. While highly speculative, we think "decision to engage" provides a biologically plausible and computationally coherent hypothesis about the neural correlates of consciousness.

1 Introduction

At the time of this writing, it seems safe to state that neuroscience has thus far failed to provide an answer to the question of how the brain gives rise to consciousness and conscious awareness. Although this essay does not provide an answer to this question, we hope it will give some indication of how to go about finding one. Our goal is to identify a framework for addressing at least some of the problems that arise. There may be a kernel of a theory of consciousness here, but we do not believe it is coherent – yet. If we are correct, and the neurobiology of decision-making is closely tied to the neurobiology of consciousness, there will be plenty of facts that will help to shape such a theory.

Our perspective brings together several ideas that might seem separate: the neurology of arousal and its disorders (Laureys 2006; Laureys et al. 2004; Posner et al. 2007), the neurobiology of decision making (Gold and Shadlen 2007), the default system (Raichle and Snyder 2007; Raichle et al. 2001), non-conscious cognition (e.g., Dijksterhuis et al. 2006; Kouider and Dehaene 2007), a bit of philosophy (Merleau-Ponty 1962), and emerging ideas about intralaminar and matrix thalamus (Jones 2001; Schiff 2010).

It may be useful to consider two broad characterizations. One is from the perspective of neurology, which tends to view the phenomenon as intrinsically linked to arousal. Consciousness refers to a state of wakefulness with organized interaction with the environment, where organized implies behavior more complex than a reflex. Consciousness is absent in sleep, coma, general anesthesia and generalized seizures. It is evanescent in stupor, perhaps less so in obtundation, and latent in the minimally conscious state. It is present – although one might say it is impaired – in confusion states with diverse forms and etiologies (delirium, toxic/infection, paraneoplastic, and psychiatric[1]). We will refer to this characterization as N-consciousness (for neurology).

The other characterization is from the philosophy of mind, which identifies a collection of mental phenomena sharing subjective, personal features. These include perceptual awareness, self-awareness, volition with awareness (i.e., authorship), a sense of free will, a sense of what it is like to be, a capacity to report narrative, introspection, and so on. Even this incomplete list portrays the daunting topic that

[1]We lack strong convictions about whether fugue states and disorders of thought, such as schizophrenia, belong in this list. We suspect these disorders will involve distinct mechanisms, which are not of the type discussed here.

makes consciousness so mysterious and special. We might argue about whether animals possess such capacities, but no one doubts that they have wakefulness and even states of confusion. We will refer to this characterization as P-consciousness (for philosophy).[2]

These conceptual perspectives lack obvious intersection, yet our central thesis is that similar neural mechanisms, computations and structures underlie many if not all of these forms of consciousness. The common feature is a decision to engage (Shadlen and Kiani 2007). Waking from sleep is a decision to engage the environment, and acting with awareness of purpose (authorship or will) involves a decision to engage a form of narrative associated with potential reportability. Thus we propose that neural mechanisms that give rise to conscious states share features with neural mechanisms that underlie simpler forms of decisions.

There is one important assumption that deserves mention at the outset. It is that a great portion of higher cognitive processing occurs without P-consciousness, that is, without awareness and without a capacity to report. We cascade actions, juggle tasks, maintain goals and highly structured cognitive sets, interact socially, navigate in artificial environments with objects that are not part of our bodies, and so on, often without the aid of P-consciousness. The topic of non-conscious processing is difficult to study, but there are some tantalizing and beautiful studies (e.g., Christoff et al. 2009; Dehaene et al. 2006; Kouider and Dehaene 2007; Lau and Passingham 2007). What seems remarkably obvious to a neurologist is that patients with disorders of higher brain function (i.e., cognitive loss) tend to miss mainly the features of their mental lives that live below the radar of consciousness. It appears that the conscious acts survive. Indeed patients exploit P-consciousness to rescue cognitive functions, much like the deliberate coping movements that are so common in the partially paralyzed.

We will return to this point, because we suspect that the neural mechanisms that give rise to decisions to engage (in certain ways) also play a role in these non-conscious functions. They too make use of decisions to engage in certain *other* ways.

2 Why View Consciousness as a Decision to Engage in a Certain Way?

There are at least two reasons to adopt this perspective. First, the formulation is already consistent with at least some aspects of both N- and P-consciousness. This is almost a matter of definition for N-consciousness. Disorders on the spectrum of coma-to-wakefulness are distinguished by a threshold for processing an external cue, such as a sound or tactile force, to engage the environment in a certain way. Reflexive withdrawal is a way that does not require consciousness. Just about

[2]Not to be confused with Ned Block's "phenomenal" consciousness (Block 2005).

anything sustained, organized and not pathologically stereotyped[3] is evidence for wakefulness, hence N-consciousness.

In sleep, our brains monitor the auditory environment for the sounds that should or should not alert us, what psychologists call a "Go vs. No-go" decision. For example, our brains choose "No-go" for unimportant sounds, like the rustling or snoring of a partner, the music or television that was on when we fell asleep, neighbors, car and train sounds, crickets at night and much birdsong at dawn. Our brains choose "Go" when a fire alarm rings, a child cries, or birds call at the right time or in some complex combination with other sounds at roughly the right time. Many a sleep-deprived mother will respond to the baby's cry but not to the tornado alarm.

Some features of P-consciousness clearly involve decision processes. Consider the spectacular demonstrations of so-called change blindness (Rensink 2000; Simons and Chabris 1999). In one famous example, a gorilla walks through a small group of students who are throwing and catching balls. Viewers who are instructed to count the throws of the students wearing white shirts, say, often fail to see the gorilla. They lack awareness of the gorilla, even when he faces the camera and beats his chest. This is not because the eye and visual cortex have failed to represent the gorilla. It is because the brain is engaged in a demanding task. A non-conscious process has decided to ignore rather than explore data in visual cortex corresponding to the gorilla. It has decided to engage the counting problem more fully. This is an example of an exploration vs. exploitation decision (Cohen et al. 2007).

Our view is that consciousness is mediated by decisions to engage, but not all decisions to engage require or imply consciousness. Again, this is based on the conviction that many sophisticated cognitive functions transpire without the aid of P-consciousness. Presumably such processes also involve exploration-exploitation decisions, to pick just one example of a decision to engage. Thus, we view consciousness as a decision to engage *in a certain way*, or set of ways. These certain ways are likely to touch on a capacity to report, to attach narrative with episodic context (e.g., before and after, place and situation). As a field, we lack an understanding of the circuits that mediate these "certain ways." But we suspect they are not qualitatively different from brain circuits involved in other behaviors, which are better understood.

The link between consciousness and decision-making provides an opportunity to ground consciousness in neurobiology. Over the past 15 years, the neural mechanisms underlying simple decisions have begun to be elucidated. These include simple perceptual decisions (Gold and Shadlen 2007; Romo and Salinas 2003; Schall 2001; Uchida et al. 2006) and value-based and social decisions (Glimcher 2003a, b; McCoy and Platt 2005; Sugrue et al. 2005). Some of the key principles are explained in the next section. The connection to decision making inspires hypotheses and guides the study of neural mechanism. It grounds the enterprise, simplifies it in some ways, and exposes deficiencies in the current emphasis on neural correlates of consciousness. That said, it offers more promise than substance.

---

[3]Certain stereotyped movements occur in persistent vegetative states, whereas others are hallmarks of epileptic seizures.

3 Neural Mechanisms of Decision Making

The remainder of this essay has a dual purpose. The first is to provide a highly selective review of some essential principles of the neuroscience of decision making – just enough detail to support the contention that there is real neuroscience here. At the very least, we wish to reassure the reader that, by tying consciousness to the neurobiology of decision-making, we are not simply relegating consciousness to another mysterious function. We do not pretend that decision making is a mature field, however. The principles and even the "facts" are not fixed, and where they appear to be so, they may be less general than we would like to believe. The second purpose is to expose tentative extensions to what is known and even more tentative connections to other areas of neuroscience. These comprise the short sections below, which describe the "intentional framework" and the problem of circuit selection.

Figures 1 and 2 show two types of tasks used in our laboratory for the study of decision making. The first is a simple perceptual decision. A monkey (or human) views a movie consisting of dynamic random dots and must decide whether the net direction of motion is to the left or right. Most of the decisions are very difficult because the stimulus consists mainly of dots appearing only briefly at random locations within the display aperture. Only a small fraction of the dots undergo displacement and then only to disappear. This fraction, termed the *percent coherence*, controls the difficulty of the discrimination. There is no actual motion to track in such a display: no dots traverse the display over extended time. The decision-maker must accumulate brief pieces of momentary evidence bearing on the two alternative hypotheses. In this sense, this simple task has more in common with cognitive decisions than with problems in perception, which rarely involve accumulation over time of independent samples of momentary evidence.

The random-dot motion task is useful for the study of decision making because so much is known about the neurobiology. Neurons in the visual cortex extract and represent the momentary evidence. Such direction-selective neurons respond to light when it is presented in a part of the visual field, termed the receptive field. The designation "direction selective" implies that the neuron responds more when motion is in one direction than in the opposite direction. Different neurons prefer different directions. The neurons that are most informative for this task are in a part of the visual cortex called MT/V5.[4] Properties of these neurons are summarized in a recent review (Born and Bradley 2005).

Four features of the MT neurons are important. (1) The intensity of the neural response, measured as a firing rate (spikes per second), is stronger when the random dot motion is in the neuron's preferred direction and when the percent coherence is greater. (2) There is a response even when the stimulus is purely random dots, without any net motion, termed 0% coherence. (3) The response itself is noisy, meaning that an estimate of the firing rate over a brief epoch, even from hundreds of

[4]MT stands for middle temporal, the name of the sulcus in the new world monkey where the area was first discovered (Allman and Kaas 1971).

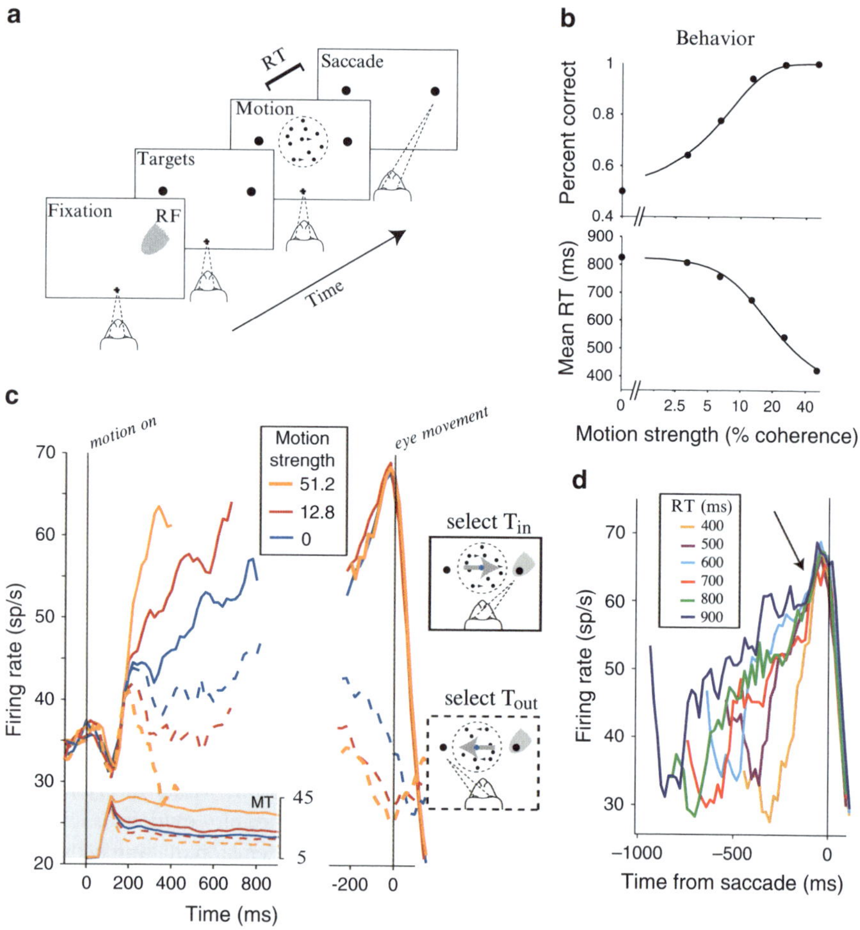

Fig. 1 Neural mechanism of a decision about direction of motion. (**a**) Choice-reaction time version of the direction discrimination task. The subject views a patch of dynamic random dots and decides the net direction of motion. The decision is indicated by an eye movement to a peripheral target. The subject controls the viewing duration by terminating each trial with an eye movement whenever ready. The *gray patch* shows the location of the response field (RF) of an LIP neuron. (**b**) Effect of stimulus difficulty on choice accuracy and decision time. Solid curves are fits of a *bounded drift-diffusion* model, which accounts simultaneously for choice and decision time. (**c**) Response of LIP neurons during decision formation. Average firing rate from 54 LIP neurons is shown for three levels of difficulty. Responses are grouped by motion strength and direction of choice, as indicated. *Left graph*: The responses are aligned to onset of random-dot motion and truncated at the median reaction time. These responses accompany decision formation. *Shaded inset* shows average responses from direction-selective neurons in area MT to motion in their preferred and anti-preferred directions (*solid* and *dashed traces*, respectively). After a transient, MT neurons respond at a nearly constant rate. The LIP firing rates approximate the integral of a difference in firing rates between MT neurons with opposite direction preferences. *Right graph*: The responses are aligned to the eye movement. For T_{in} choices (*solid curves*), all trials reach a stereotyped firing rate before saccade initiation. We think this level represents a threshold or bound, which is sensed by other brain regions to terminate the decision. (**d**) Responses grouped by reaction time. Only T_{in} choices are shown. *Arrow* shows that the stereotyped firing rate occurs ~70 ms before saccade initiation (adapted with permission from Gold and Shadlen 2007; Roitman and Shadlen 2002; Shadlen et al. 2006)

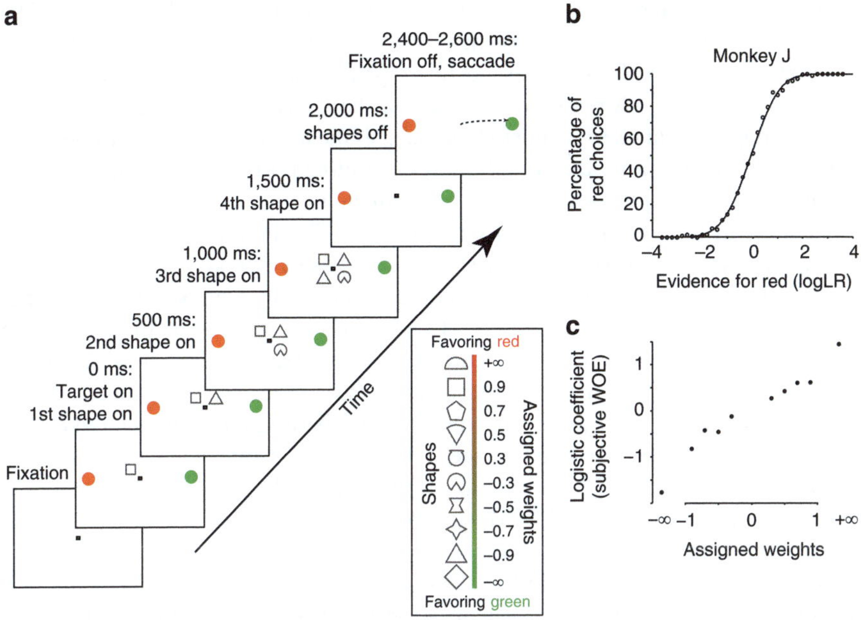

Fig. 2 Probabilistic categorization task. (**a**) Task sequence. Four shapes are presented sequentially on the computer monitor near the center of gaze. After a brief delay period, the monkey makes an eye movement (saccade) to either the *red* or *green* choice target. During neural recording, one of the choice targets was in the response field of the neuron. The shapes are selected randomly in each trial from a larger set of ten (*inset*). The reward is determined probabilistically by summing the weights associated with the four shapes. The sum is the logarithm of the odds that the *red* target will be the one rewarded on that trial. (**b**) Performance. The fraction of *red choices* is plotted as a function of the logLR conferred by the four shapes in favor of red. *Curves* are logistic fits to the data. Only trials that have probabilistic reward (P≠1 or 0) are included in this graph. (**c**) Effect of individual shapes on choice. The leverage of each of the *ten shapes* on the probability of a *red choice* was inferred using logistic regression, which is the contribution that the shape has on the log10 of the odds of a red choice. These values are plotted as a function of the assigned weights. Movie example trials from the experiment, along with neural recordings, can be viewed at http://www.nature.com/nature/journal/v447/n7148/suppinfo/nature05852.html (adapted with permission from Yang and Shadlen 2007)

neurons, is highly variable. (4) The neurons respond to the visual information with short latency and they stop responding when the stimulus is not present. The importance of features 1 and 2 is that the evidence these neurons provide for decision making is graded. The importance of feature 3 is that the evidence is unreliable. Indeed it is possible for the left-preferring neurons to respond more than the right-preferring neurons, even when the motion is rightward, and thus lead to an error of perception. The fourth feature indicates that there is no build up or accumulation and no memory of the past. When the motion is on, the neuron responds at a constant (albeit noisy) rate (see Fig. 1c, gray inset). These are neurons that keep up with a changing world.

In short, the MT neurons supply the momentary evidence to the decision. The better the evidence, the more likely the decision will be correct and the faster it will complete. But the MT responses do not represent the decision outcome or its formation. They do not represent the state of the decision once it is made or the accumulation of evidence leading to this state.[5] If the monkey must hold its decision in working memory after the stimulus has been turned off, the MT neurons are no longer informative, yet the decisions are unaffected.

Neurons in the association cortex are different. Their responses can linger for seconds, even tens of seconds. Like visual neurons, they respond only under the right conditions, and the intensity of their firing rates represent an amount of something, but it is often difficult to say what. That depends on what kind of information they receive and to what kinds of neurons they project. For neurons in the lateral intraparietal cortex (area LIP), the input is from visual cortex and the output is mainly to structures that control eye movements or the focus of spatial attention (Andersen 1995; Andersen and Buneo 2002; Bracewell et al. 1996; Colby and Goldberg 1999; Mazzoni et al. 1996). LIP neurons are well positioned to convert the stream of momentary evidence, ascribed to MT neurons, into a quantity that is used to make the decision—termed a decision variable—and to represent its outcome.

Neurons in LIP have spatially selective persistent neural activity. Like MT neurons, there are restricted regions of the visual field that support responses and other regions that do not. We use the term *response field* (instead of receptive field) because the LIP response is affected both by visual targets and a plan to make an eye movement or shift attention to that location. In contrast to MT, the responses of LIP neurons persist in the absence of continuous visual stimulation. And while they are associated with a plan to make an eye movement, they do not obligate an immediate movement.

We believe such selective persistent activity holds the key to understanding higher cognitive function, its emergence in evolution and its impairment in disease, because our most cherished mental functions require that neural computations can transpire a time frame that is (1) not governed by immediate change in the environment and (2) not tied to immediate change in body musculature. Such *freedom from immediacy* probably arose as the cortical mantle expanded, thereby inserting contingency into the basic sensory-motor design. We will return to this theme below.

LIP neurons contribute to decision formation on the task illustrated in Fig. 1a. During decision formation, the firing rate of these neurons represents the accumulation of momentary evidence in the visual cortex. As shown in Fig. 1c, the accumulation rises or falls depending on the direction and strength of the motion. These averages belie the tremendous variability in firing rate from trial to trial. On a single trial, the firing rate would resemble the path of a particle undergoing drift-diffusion or biased random walk. On each trial, the LIP firing

[5] We are ignoring the small modulation of sensory neurons associated with choice (Britten et al. 1996; Celebrini and Newsome 1994). In some settings, part of this modulation could represent the outcome of the choice (Krug 2004; Nienborg and Cumming 2009).

rate represents the running accumulation of the momentary evidence they receive from the neurons in MT. Since this momentary evidence is variable, the accumulation resembles Brownian motion of a particle, hence the analogy to diffusion and random walks (Churchland et al. 2011).

When the firing rates of LIP neurons reach a critical level, the decision process halts (Fig. 1d). If the task is set up to study both choice and reaction time, then the stopping time results in a behavioral outcome – the reaction time (Fig. 1b). If the stimulus is presented for a long time, or if the experimenter imposes a delay period after turning off the motion stimulus (i.e., a memory-delay), then there is no direct behavioral manifestation of the stopping, but we have shown that it occurs (Kiani et al. 2008). The brain reaches a commitment and ignores additional evidence.

If, during the delay period, the monkey is allowed an opportunity to opt out of the left-right task to obtain a small but certain reward, it does so when it is less likely to choose the correct motion direction. Even for repetitions of identical stimuli, monkeys opt out when they are more likely to make an error. Importantly, the firing rate of LIP neurons is nearer the neutral level when the monkey chooses this low confidence option. The firing rates encode a degree of confidence, that is, the probability of answering correctly (Kiani and Shadlen 2009). These and other observations suggest that the firing rates of neurons like those in LIP do not represent quantities associated with stimuli or actions per se, but are best articulated using the language of probability, value, utility, and costs – terms associated with decision theory.

Such neural computations are key components of any conscious decision or plan of action, although we can say little about the monkeys' state of awareness in our experiments. The confidence study shows that the brain computes quantities that support a form of meta-cognitive reasoning about the degree of certainty in a decision, but the observation does not imply that the monkey is aware of the process (but see Hampton 2001; Kornell et al. 2007; Smith et al. 2003). That said, the process that transpires once the sure-target option is made available resembles a decision to report about the state of another mental process – here, the decision that occurred a second or so ago during stimulus viewing. Again, we do not believe it is possible to ascertain whether this decision to engage the sure-target in this way is associated with awareness. A more rewarding speculation is that the mechanism exposed in these experiments resembles the ones in our own brains when we engage in a way that we experience as being consciously aware. Using this idea, we can exploit animal models of cognition to study the neural mechanisms of human consciousness. We need not resolve the question of "what it is like" to be a monkey (Nagel 1974).

Another experimental observation that supports the representation of probabilities by LIP neurons comes from the probabilistic categorization task shown in Fig. 2 (Yang and Shadlen 2007). In this task, the monkey must decide between a red and a green target. One or the other will render a reward on a random half of trials. On any one trial, however, the probability of "reward at red" is governed by a set of four shapes, shown sequentially on the video monitor. These shapes are drawn randomly (with replacement) from a larger set of ten, half of which support "reward at red" by varying degrees. The other half support "reward at green" by the same varying degrees. The experiment tests whether the monkey can reason from the probabilistic evidence and make the better choice, based on the four random shapes shown on any one trial. This task is quite challenging. Although there is a better choice on any one trial, there is no guarantee that selecting it will lead to reward, because the reward is ultimately rendered probabilistically. After training, however, monkeys learn this task (Fig. 2b) and that some shapes are more reliable predictors than others (Fig. 2c).

The neural responses from this task are best appreciated by viewing movies from the experiment.[6] They show the same type of LIP neuron studied in the motion experiment. Here, the red or the green choice target is in the neuron's response field. What is immediately evident in the movies is that the neuron performs a running sum of the positive and negative support from the shapes. A more quantitative analysis reveals that the firing rate is proportional to the logarithm of the ratio of probabilities (the log-likelihood ratio or log-odds ratio), which is an intuitive solution. If the probabilities are equal, then the ratio is 1 and $\log(1) = 0$. If the probability favors the target in the response field, then the ratio exceeds 1 and the log is positive, whereas if the probability favors the other target, then the ratio is less than one and the log is negative. Moreover, if the neural response represents a logarithm of a probability ratio based on the first shape shown in the trials, then when a second shape arrives, it is sensible to simply increment or decrement the response by the new log-likelihood ratio. That is what you can hear in the audio track of the movies.

This brief survey exposes four principles of neural function in the service of decision-making. (1) The response can evolve gradually in time to represent the accumulation of evidence from multiple sources in time. (2) The mechanism includes a termination rule, that is, a criterion for finishing the decision. This might be based on the amount of evidence, the passage of time as in a deadline, or a computation involving value, costs and so on. (3) The computations bear resemblance to probabilistic inference. The general importance here is that it allows neurons to use spike rate to represent intensities that are loosely coupled to "degree of belief in..." or "likelihood that..." or "expected loss if...." (4) The neurons associated with decision formation defy classification as sensory or motor. They are influenced by sensory stimuli and their responses probably influence motor function, but they respond without sensory input and they do not obligate a movement. They lie at the nexus of sensory processing and motor planning.

4 Intentional Framework and Circuit Selection

The fourth principle brings us back to the main focus of the essay. It inspires most if not all of the studies of decision-making in our lab, and it helps us see beyond this one brain area. We assume that area LIP is not particularly special. We observe

[6] http://www.nature.com/nature/journal/v447/n7148/suppinfo/nature05852.html

decision-related activity in LIP because the monkeys are trained to communicate their decisions with an eye movement. LIP is well suited for decision making in these tasks because it receives input from visual cortex and because its main output targets are structures that control eye movements. Unless LIP is special, it seems likely that decision formation will be evident in other structures that are tied to intention or planned actions, or more generally to the selection of neural circuitry that controls a behavior. We refer to this architecture as an intentional framework, to contrast it with the more representational framework (Shadlen et al. 2008).

There is some support for this idea. The parietal areas that neighbor LIP receive mainly visual information, but they target premotor cortical areas devoted to other motor functions. For example, the medial intraparietal area (MIP) projects to regions that control reaches to targets, and the anterior intraparietal area (AIP) projects to regions that control hand posture during grasp. It is tempting to speculate that these areas do something very much like LIP. They operate on the stream of activity from the visual cortex and construe it as evidence in support of a proposition. For LIP, the proposition is not really about direction of motion but ultimately about which eye movement to make. Indeed, it is already known that MIP behaves similarly to LIP when the monkey communicates its decision by touching a spot on a touch-screen device (Andersen and Buneo 2002; de Lafuente et al. 2009; Scherberger and Andersen 2007; Snyder et al. 2000). For AIP, the stream of activity bears on geometry (Janssen et al. 2008), but it is perhaps more apt to describe the responses in terms of a possible grasp posture (Cisek 2007).

An obvious shortcoming of the intentional view is that it would appear to relegate all of ideation to motor planning. A natural objection to the preceding paragraphs is as follows. The monkey may be deciding about where to move its eyes, but I, as a human, decide about the motion, independently of what action or word I would use to communicate my decision, indeed whether I communicate it at all.

We agree that ideation does not necessitate action. Nonetheless, we believe the essential features of abstract ideation are visible in the sensory-motor decision mechanisms we study. Figure 3 illustrates a simple "abstract" decision. The subject decides the direction of motion without knowledge of the motor response required to indicate an answer. In the epoch during motion viewing and the onset of the colored choice targets, the monkey forms and remembers a decision about motion, not about the next action. Neural circuits devoted to planning a particular action do not reflect accumulating evidence (Gold and Shadlen 2003, 2007). Instead, the brain must decide on a plan to make another decision: to make an eye movement to one or another target based on color. The key to abstraction is to view decision making not as information bearing on an action but on the selection of circuitry that mediates another decision.

Indeed, the larger cortical mantle might support higher cognitive function in humans by allowing us to make decisions about decisions about decisions about . . . decisions to do something. It is not hard to see the elements of symbol manipulation in the example of Fig. 3. By cascading association areas that do not project to motor structures but instead to other association areas, we can imagine the basis of much more complexity. For example, we can appreciate the layers in performing an

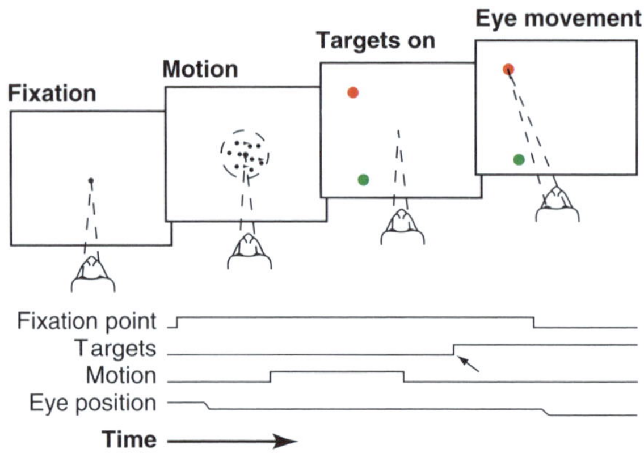

Fig. 3 A direction discrimination task used to separate abstract decision making from motor planning. The subject must decide the net direction of dynamic random dot motion: *left* or *right*. Decision formation occurs before the targets are displayed at unpredictable locations. The monkey chooses *red* (or *green*) for *right* (or *left*). Before the targets are turned on (*small arrow*), the monkey must commit to a proposition that is more abstract than a plan to move the eyes (adapted with permission from Gold and Shadlen 2003)

action to achieve a goal, in imitation of another's actions, in construing from another's action the goal that led to the other's action, in mirroring this goal – steps toward a neurobiology of "theory of mind."

The intentional framework has much in its favor. This is not the place to expound all of its virtues, but as this meeting takes place in Paris, it is a pleasure to mention that the framework was anticipated in the writings of Maurice Merleau-Ponty. He regarded vision not as inference on visual impressions (representation) but as answers to more purposeful interrogations of the environment (Merleau-Ponty 1962). The panorama to the blind man is not the sequence of vibrations from the stick to the hand but answers to questions about obstacles, the ground surface, steps, cliffs, walls and so on.

Much of the motivation behind many mainstream theories of perception and consciousness in neuroscience is driven by over-enthusiasm for the representational framework. It leads to bizarre solutions that elevate agnostic representations of information to the status of perception and awareness by oscillating it, synchronizing it or enhancing its power spectrum in some frequency band.[7] Even if such measurements were to occur reliably with the phenomena they supposedly explain, by what mechanism do they arise? What brain structure *decides* to wiggle some part of the representation and thus render it available for conscious awareness?

[7]The change blindness demonstrations underscore the inadequacy of neural representation to explain perception. The unseen object (e.g., the gorilla among the ball players) is represented in visual cortex with greater amplitude than many low-contrast but highly visible objects.

5 Circuit Selection and Configuration

The action of neurons in LIP is not to move body parts but to influence other neurons in the brain. LIP influences eye movements by guiding the selection of circuits in the frontal eye field and superior colliculus. Some other brain structure probably makes a decision that the context is befitting and selects the appropriate LIP circuits to construe evidence from visual cortex as bearing on the salience of potential saccade targets (and not other items present in the visual field). Indeed the circuit must also be configured so that the evidence is compared in a sensible way (e.g., rightward direction sensors from part A of the visual field provide support for LIP neurons that represent part B of the visual field).[8]

The mechanism underlying circuit selection is unknown. We believe it is among the most important problems in systems neuroscience, in part because of its connection to the neurobiology of consciousness. A decision to engage is simply that: evaluation of evidence leading to turning on another circuit and configuring the flow of information – for example, so that the new spot of activity in the primary visual cortex gets inspected and identified as a gorilla (or not).

The idea is not that all such operations lead to conscious awareness but rather that the ones that do so use this mechanism to engage particular circuits. A wide variety of cognitive functions requires that one operation, naturally construed as a decision, leads to the activation of one or more of a larger set of circuits. We believe this can occur without consciousness and does not necessarily lead to conscious awareness, as in the perceptual decision tasks, when contextual information induces clusters of neurons in LIP to represent the integrated evidence from the visual cortex. In this view, the N-conscious processes are those that allow the brain to make more non-conscious decisions about what else to engage, whereas the P-conscious processes are decisions to engage in certain ways, most prominently for communicating, pointing to another – that is to say, reporting.

6 Some Candidate Structures for Decisions to Engage

Several brain structures emerge as potential targets of inquiry. When we consider areas of the brain that are involved in arousal, the midbrain reticular formation and the intralaminar nuclei are at the top of the list. These are the sites that Schiff and colleagues targeted in their successful restoration of consciousness to a severely impaired patient (Schiff 2010; Schiff et al. 2007). It is now recognized that intralaminar nuclei make different types of connections with the cortex than the rest of the thalamus. Instead of targeting layer IV, they target superficial layers and tend to arborize more extensively. This pattern is also evident in a class of neurons

[8]For present purposes, we are lumping circuit selection and configuration. It is the former that has the clearer connection to decisions to engage.

outside the intralaminar nuclei that share certain molecular markers. These neurons and the intralaminar nuclei comprise the thalamic matrix (Jones 2001). One intriguing idea is that these matrix neurons play a role in cortical circuit selection. Support for this view might be adduced from a recent paper from M. Sherman and colleagues (Theyel et al. 2009).

Other cortical areas are likely to play a role in circuit selection. As mentioned above, such decisions about *what to make decisions about* need not invoke P-consciousness. Presumably executive control arises in a variety of contexts that contribute to both conscious and nonconscious cognition (e.g., Del Cul et al. 2009; Miller 2000). Some cortical areas that are part of a default network, however, seem to play a role in monitoring the world exactly when we are not engaged (Christoff et al. 2009; Raichle and Snyder 2007; Raichle et al. 2001). They seem poised to make decisions about whether or not to engage, and perhaps it is a large enough system to support decisions to engage in certain ways. One consideration that plays a role in our thinking about cortical control is the constraint on wiring. It simply is not the case that a central structure can address the entire cortex, and we do not know of a switchboard in the brain, although matrix thalamus comes closest.

Language areas seem like obvious candidate structures for decisions to engage in a narrative way. Such areas might be targets of circuit selection or they might participate in decisions to engage for possible reporting or attaching narrative. A more inchoate expression of such functionality might reside in the association auditory cortex. Consider the prominence of audition in the examples mentioned earlier concerning arousal! It is intriguing that the evolution of auditory association cortex might parallel the capacities we invoke when we engage in the ways that touch on P-consciousness: attaching narrative, episodic/declarative context, pointing out a visual object to another individual.

The following coincidence may be of interest to some readers. Rhesus monkeys are notoriously difficult to train on high-level auditory tasks. They can localize and recognize calls, but we have had little success training them to make the kinds of arbitrary associations that are exemplified in the tasks described above. A monkey can make an eye movement to the location of a hidden speaker, but we have found it difficult to train monkeys to decide about the duration of intervals between clicks and indicate a choice with an eye movement.[9] There may be many reasons for this, including our own ineptitude, but many researchers in the field are struck by the lack of auditory aptitude of macaques (Fritz et al. 2005). Coincidentally, monkeys seem to possess little parietal cortex devoted primarily to audition. Earlier, we mentioned three areas along the intraparietal sulcus that are devoted to vision-action associations but can be co-opted for more abstract visual perceptual decisions. These are somatosensory areas rostral to the sulcus. But as one moves toward the auditory cortex, the sulcus disappears. There is auditory input to parietal regions that are classically identified as visual association areas, whereas the remaining regions, near the temporo-parietal junction, project to the most rostral

[9]We have published and are now conducting several timing tasks, which are easier for humans when using sounds. Monkeys are far more proficient using intervals between visual cues.

pole of the prefrontal cortex (Kaas and Hackett 2000; Kaas et al. 1999; Poremba and Mishkin 2007; Poremba et al. 2003).

If the evolution of auditory association cortex remains delayed in the hominid branch of *primata*, maybe we can begin to recognize why language developed late and only in one hemisphere. And since there is no way to close our ears, perhaps we had to have more elaborate systems of decisions to engage based on stimulation. Perhaps our brains exploit this capacity to engage other systems. These are a few highly speculative justifications for the view that auditory association cortex and its thalamic targets might play a privileged role in decisions to engage. This is obviously not the only path to P-consciousness, but it may be one that is highly developed. And this is not to say that hearing must be intact to use this neural substrate (e.g., in deafness).

7 Limitations and Dividends of the Idea

The main shortcomings of the decision to engage idea are (1) it lacks a known mechanism, (2) it is probably involved in non-conscious as well as conscious processes, (3) its parallel "intentional" architecture fails to capture the unity of experience captured by conscious awareness, and (4) it does not explain the subjective "what it is like to be..." aspects of consciousness, including *qualia*. This is not a complete list, but it is enough to have a stab at.

(1) The mysterious step in the "decision to engage" is the mechanism underlying circuit selection. It is poorly understood in the simplest of perceptual decisions. That said, it is a mechanism, and the notion that it can be studied in simpler contexts is a dividend of the theory. We do not know how LIP "selects" neurons in the frontal eye field or superior colliculus, when a decision about motion leads to an eye movement, but the question is tractable. A decision to engage links neural mechanisms mediating N- and P-consciousness to the neurobiology of decision making. Put more bluntly, it is a testable idea about a mechanism. It does not assert that a mysterious function adds an incantation – be it synchrony or oscillations or power in a range of frequencies – to a representation, thus rendering it available to awareness. Even if rendering to awareness were marked by such an "incantation," one would ask about the mechanisms in the brain that led to its application. If the answer is, "a decision to activate via matrix thalamus a set of cortical circuits that decide to use vision," we would recognize a mechanism with similarities to others we are beginning to understand.

(2) Admittedly, decisions to engage are also shared by the organized behaviors that neither require nor reach the level of consciousness. We speculate that the difference is partially due to the brain structures that initiate the decision to engage (and perhaps whether such initiation is programmed by a conscious or unconscious mental process). Regardless of how the process is initiated, the

shared mechanism of decision-to-engage provides an opportunity to study consciousness. If we can understand how a decision about a simple sensory stimulus (e.g., noisy motion) leads to the selection of a motor circuit (e.g., in the frontal eye field and/or superior colliculus to plan or initiate an eye movement), we will have taken a step toward understanding a mechanism that could also engage other circuits that lead to exploration, pointing out to another being, communicating and attaching narrative.

(3) Decisions to engage in particular ways conform to a brain organization that lacks a central executive or global workspace with access to all functions. This may be limiting, or simply wrong, were it to turn out that such a central, integrative, organizing structure exists. Yet, we cannot think of a brain structure that could qualify as such a structure. Brain regions that project broadly (e.g., the locus coeruleus) lack the computational capacity – or even representational capacity – to serve as a central workspace for consciousness. We do not pretend to understand how a parallel architecture can support the unity of our experience, but we suspect it can be accounted for by a parallel organization viewed in the intentional framework. If perception is not so much a declaration of labels but answers to questions, posed serially or in parallel, about a place or time or relationship of places and times, then those answers have unity arising from the "aboutness" of the intention. Von Helmholtz held such an intentional view of space perception (von Helmholtz 1925). Merleau-Ponty expresses this view in the Phenomenology of Perception (Merleau-Ponty 1962).

(4) The mechanism alone fails to address the distinction between the personal, subjective aspects of consciousness – what is it like to be me; how do I experience red, love, sorrow, etc.; ownership of my feelings, perceptions and acts – from other less subjective aspects of awareness and agency. We surmise that these are different ways of engaging that involve different brain structures. We suspect that different ways of engaging the world have attributes that can facilitate or prevent subjective narratives and/or qualia. Qualia might arise when we engage in ways that involve social and communicative components, attach narrative context with declarative attributes (e.g., before, after, where, what else is present) and sensorimotor contingencies (O'Regan and Noë 2001). The subjective aspects of conscious experience are multi-faceted and complex, hence unlikely to be explained by a single process. The decision to engage is the first building block of a subjective conscious experience, not the entirety of it.

The main dividend to the notion of consciousness as a decision to engage is that it offers a glimpse of an actual neural mechanism that can be studied in a variety of contexts and model systems. If we wake up to sounds through a decision to engage, or if we decide to engage a distant object with a decision to report, say, via pointing to another, we can bring to bear what is already known about the neurobiology of perceptual decisions. We can ask whether the bounded accumulation of evidence in the posterior cingulate cortex (PCC), to choose one interesting example, leads to the activation of another cortical area, via intralaminar and matrix thalamus, and whether this occurs once a threshold or bound is achieved by the firing rate of neurons in PCC.

Such consideration of putative mechanism contrasts with the traditional quest for a neural correlate of consciousness (NCC), which may well mark the presence of consciousness but which fails to explain its derivation. That said, our hypothesis is compatible with popular ideas about the NCC. If it so happens that power in a range of frequencies measured in local field potentials is a signature of conscious processing, then one ought to ask what is the mechanism that causes this physiological change. It could be the case that when matrix thalamus activates cortex, it leads to changes in the cortical microcircuit that are associated with increased power in the gamma band of frequencies measured in the LFP. We are not invested in this view but mention it as testimony to the compatibility of "decisions to engage" with existing theories and observations.

8 Concluding Remarks

The concept of a decision to engage links the neurobiology of consciousness to the field of decision making. It has the virtue of tying together characterizations of consciousness employed in clinical neurology with the phenomenology that we associate with the mind's most precious pursuits. It may guide future experiments and, if correct, it would render broad areas of systems, cellular and molecular neuroscience relevant to the study of consciousness.

References

Allman JM, Kaas JH (1971) A representation of the visual field in the caudal third of the middle temporal gyrus of the owl monkey (*Aotus trivirgatus*). Brain Res 31:85–105

Andersen RA (1995) Encoding of intention and spatial location in the posterior parietal cortex. Cereb Cortex 5:457–469

Andersen RA, Buneo CA (2002) Intentional maps in posterior parietal cortex. Annu Rev Neurosci 25:189–220

Block N (2005) Two neural correlates of consciousness. Trends Cogn Sci 9:46–52

Born RT, Bradley DC (2005) Structure and function of visual area MT. Annu Rev Neurosci 28:157–89

Bracewell RM, Mazzoni P, Barash S, Andersen RA (1996) Motor intention activity in the macaque's lateral intraparietal area. II. Changes of motor plan. J Neurophysiol 76:1457–1464

Britten KH, Newsome WT, Shadlen MN, Celebrini S, Movshon JA (1996) A relationship between behavioral choice and the visual responses of neurons in macaque mt. Vis Neurosci 13:87–100

Celebrini S, Newsome WT (1994) Neuronal and psychophysical sensitivity to motion signals in extrastriate area mst of the macaque monkey. J Neurosci 14:4109–4124

Christoff K, Gordon AM, Smallwood J, Smith R, Schooler JW (2009) Experience sampling during fMRI reveals default network and executive system contributions to mind wandering. Proc Natl Acad Sci USA 106:8719–8724

Churchland AK, Kiani R, Chaudhuri R, Wang XJ, Pouget A, Shadlen MN (2011) Variance as a Signature of Neural Computations during Decision Making. Neuron 69:818–831
Cisek P (2007) Cortical mechanisms of action selection: the affordance competition hypothesis. Phil Trans R Soc B Biol Sci 362:1585–1599
Cohen JD, McClure SM, Yu AJ (2007) Should I stay or should I go? How the human brain manages the trade-off between exploitation and exploration. Phil Trans R Soc B Biol Sci 362: 933–942
Colby CL, Goldberg ME (1999) Space and attention in parietal cortex. Annu Rev Neurosci 22: 319–349
de Lafuente V, Jazayeri M, Shadlen MN (2009) Decision related activity in areas lip and mip during reaches and saccades. Neurosci Mtg Planner, vol 652.9. Society for Neuroscience, Chicago, IL
Dehaene S, Changeux JP, Naccache L, Sackur J, Sergent C (2006) Conscious, preconscious, and subliminal processing: a testable taxonomy. Trends Cogn Sci 10:204–211
Del Cul A, Dehaene S, Reyes P, Bravo E, Slachevsky A (2009) Causal role of prefrontal cortex in the threshold for access to consciousness. Brain 132:2531–2540
Dijksterhuis A, Bos MW, Nordgren LF, van Baaren RB (2006) On making the right choice: the deliberation-without-attention effect. Science 311:1005–1007
Fritz J, Mishkin M, Saunders RC (2005) In search of an auditory engram. Proc Natl Acad Sci USA 102:9359–9364
Glimcher P (2003a) The neurobiology of visual-saccadic decision making. Annu Rev Neurosci 26:133–179
Glimcher P (2003b) Decisions, uncertainty, and the brain: the science of neuroeconomics. MIT, Cambridge, MA
Gold JI, Shadlen MN (2003) The influence of behavioral context on the representation of a perceptual decision in developing oculomotor commands. J Neurosci 23:632–651
Gold JI, Shadlen MN (2007) The neural basis of decision making. Annu Rev Neurosci 30: 535–574
Hampton RR (2001) Rhesus monkeys know when they remember. Proc Natl Acad Sci USA 98:5359–5362
Janssen P, Srivastava S, Ombelet S, Orban GA (2008) Coding of shape and position in macaque lateral intraparietal area. J Neurosci 28:6679–6690
Jones EG (2001) The thalamic matrix and thalamocortical synchrony. Trends Neurosci 24:595–601
Kaas JH, Hackett TA (2000) Subdivisions of auditory cortex and processing streams in primates. Proc Natl Acad Sci USA 97:11793–11799
Kaas JH, Hackett TA, Tramo MJ (1999) Auditory processing in primate cerebral cortex. Curr Opin Neurobiol 9:164–170
Kiani R, Shadlen MN (2009) Representation of confidence associated with a decision by neurons in the parietal cortex. Science 324:759–764
Kiani R, Hanks TD, Shadlen MN (2008) Bounded integration in parietal cortex underlies decisions even when viewing duration is dictated by the environment. J Neurosci 28:3017–3029
Kornell N, Son LK, Terrace HS (2007) Transfer of metacognitive skills and hint seeking in monkeys. Psychol Sci 18:64–71
Kouider S, Dehaene S (2007) Levels of processing during non-conscious perception: a critical review of visual masking. Phil Trans R Soc Lond B Biol Sci 362:857–875
Krug K (2004) A common neuronal code for perceptual processes in visual cortex? Comparing choice and attentional correlates in V5/MT. Phil Trans R Soc Lond B Biol Sci 359:929–941
Lau HC, Passingham RE (2007) Unconscious activation of the cognitive control system in the human prefrontal cortex. J Neurosci 27:5805–5811
Laureys S (2006) Tracking the recovery of consciousness from coma. J Clin Invest 116:1823–1825
Laureys S, Owen AM, Schiff ND (2004) Brain function in coma, vegetative state, and related disorders. Lancet Neurol 3:537–546

Mazzoni P, Bracewell RM, Barash S, Andersen RA (1996) Motor intention activity in the macaque's lateral intraparietal area. I. Dissociation of motor plan from sensory mechanisms and behavioral modulations. J Neurophysiol 76:1439–1456
McCoy AN, Platt ML (2005) Expectations and outcomes: decision-making in the primate brain. J Comp Physiol A Neuroethol Sens Neural Behav Physiol 191:201–211
Merleau-Ponty M (1962) Phenomenology of perception. Routledge and Kegan Paul, London
Miller EK (2000) The prefrontal cortex and cognitive control. Nat Rev Neurosci 1:59–65
Nagel T (1974) What is it like to be a bat? Phil Rev 83:435–450
Nienborg H, Cumming BG (2009) Decision-related activity in sensory neurons reflects more than a neuron's causal effect. Nature 459:89
O'Regan JK, Noë A (2001) A sensorimotor account of vision and visual consciousness. Behav Brain Sci 24:939–973
Poremba A, Mishkin M (2007) Exploring the extent and function of higher-order auditory cortex in rhesus monkeys. Hear Res 229:14–23
Poremba A, Saunders RC, Crane AM, Cook M, Sokoloff L, Mishkin M (2003) Functional mapping of the primate auditory system. Science 299:568–572
Posner JB, Saper CB, Schiff ND, Plum F (2007) Plum and Posner's diagnosis of stupor and coma. Oxford University Press, New York
Raichle ME, Snyder AZ (2007) A default mode of brain function: a brief history of an evolving idea. Neuroimage 37:1083–1090
Raichle ME, MacLeod AM, Snyder AZ, Powers WJ, Gusnard DA, Shulman GL (2001) A default mode of brain function. Proc Natl Acad Sci USA 98:676–682
Rensink RA (2000) Seeing, sensing, and scrutinizing. Vis Res 40:1469–1488
Roitman JD, Shadlen MN (2002) Response of neurons in the lateral intraparietal area during a combined visual discrimination reaction time task. J Neurosci 22:9475–9489
Romo R, Salinas E (2003) Flutter discrimination: neural codes, perception, memory and decision making. Nat Rev Neurosci 4:203–218
Schall JD (2001) Neural basis of deciding, choosing and acting. Nat Rev Neurosci 2:33–42
Scherberger H, Andersen RA (2007) Target selection signals for arm reaching in the posterior parietal cortex. J Neurosci 27:2001–2012
Schiff ND (2010) Recovery of consciousness after brain injury: a mesocircuit hypothesis. Trends Neurosci 33:1–9
Schiff ND, Giacino JT, Kalmar K, Victor JD, Baker K, Gerber M, Fritz B, Eisenberg B, Biondi T, O'Connor J, Kobylarz EJ, Farris S, Machado A, McCagg C, Plum F, Fins JJ, Rezai AR (2007) Behavioural improvements with thalamic stimulation after severe traumatic brain injury. Nature 448:600–603
Shadlen MN, Kiani R (2007) Neurology: an awakening. Nature 448:539–540
Shadlen MN, Hanks TD, Churchland AK, Kiani R, Yang T (2006) The speed and accuracy of a simple perceptual decision: a mathematical primer. In: Doya K, Ishii S, Rao R, Pouget A (eds) Bayesian brain: probabilistic approaches to neural coding. MIT, Cambridge, pp 209–237
Shadlen MN, Kiani R, Hanks TD, Churchland AK (2008) Neurobiology of decision making: an intentional framework. In: Engel C, Singer W (eds) Better than conscious? Decision making, the human mind, and implications for institutions. MIT, Cambridge, pp 71–102
Simons DJ, Chabris CF (1999) Gorillas in our midst: sustained inattentional blindness for dynamic events. Perception 28:1059–1074
Smith JD, Shields WE, Washburn DA (2003) The comparative psychology of uncertainty monitoring and metacognition. Behav Brain Sci 26:317–339, discussion 340–373
Snyder LH, Batista AP, Andersen RA (2000) Intention-related activity in the posterior parietal cortex: a review. Vis Res 40:1433–1442
Sugrue LP, Corrado GS, Newsome WT (2005) Choosing the greater of two goods: neural currencies for valuation and decision making. Nat Rev Neurosci 6:363–375

Theyel BB, Llano DA, Sherman SM (2009) The corticothalamocortical circuit drives higher-order cortex in the mouse. Nat Neurosci 13:84–88
Uchida N, Kepecs A, Mainen ZF (2006) Seeing at a glance, smelling in a whiff: rapid forms of perceptual decision making. Nat Rev Neurosci 7:485–491
von Helmholtz HLF (1925) Treatise on physiological optics. Dover, New York
Yang T, Shadlen MN (2007) Probabilistic reasoning by neurons. Nature 447:1075–1080

Thinking About Brain and Consciousness

Antonio Damasio

Abstract In this brief article I present as a preamble to a developing theory a summary of ideas and evidence discussed elsewhere (see Damasio and Meyer (2008) In: Laureys S, Tononi G (eds) The neurology of consciousness. Elsevier, pp 3–14; Damasio (2010) Self comes to mind. Pantheon). They pertain to the nature of consciousness, to the position and role of consciousness in evolutionary history, and to how the brain constructs consciousness at the level of large-scale systems.

1 Defining Consciousness

I define normal or standard consciousness as a mind state in which the regular flow of mental images has been enriched by subjectivity, that is, by a reference to the organism within which the mind is being constructed. That reference, which I designate as the self process, is also built with mental images that represent ongoing body states. Because self images have a privileged connection to the body, they are "felt". The resulting conscious mind is a natural, ongoing display of facts and events occurring in the organism's surround, relative to felt facts and events occurring in the organism's interior. The self process, which can be simple or complex, offers to the conscious mind the possibility of experiencing the organism's existence and that of the world around.

I suggest that a plain mind process devoid of self elements is not likely to provide "experience". Experience requires an experiencer, the provider of subjectivity, and that is what the self process, even if at the simplest levels, confers upon a plain mind.

A. Damasio (✉)
Brain and Creativity Institute, University of Southern California, Hedco Neuroscience Building Room 126, 3641 Watt Way, Los Angeles, CA 90089-2520, USA
e-mail: damasio@college.usc.edu

S. Dehaene and Y. Christen (eds.), *Characterizing Consciousness: From Cognition to the Clinic?* Research and Perspectives in Neurosciences,
DOI 10.1007/978-3-642-18015-6_3, © Springer-Verlag Berlin Heidelberg 2011

2 Constructing the Conscious Mind

Body signals are a foundation for the conscious mind. The most stable aspects of body function are represented in the brain, in the form of maps, and constitute the substrates for the varied sensory images that we experience in the mind (visual, auditory, etc.). I hypothesize that the mental images of the body produced by the body-mapping structures constitute the "protoself", the lowest level of the self process. The basic body-mapping and image-making structures are located below the level of the cerebral cortex, in a region known as the upper brainstem, a phylogenetically old part of the brain shared with many other species.

The brain's protoself structures are not merely *about* the body: they are inseparably connected to the body. Varied parts of the body continuously signal the brain with their signals and are signaled back by the brain. I posit that this arrangement creates a perpetual resonant loop, broken only by brain disease. The arrangement has a physiological purpose: by representing the conditions of the body in the brain, it allows the brain to respond to those conditions and regulate life according to homeostatic needs. As a result, the protoself structures have a privileged and direct relationship to the body. The images they engender regarding the body are conceived in circumstances different from those of other brain images. I regard this privileged position as a clue to why such images are felt. Body feeling is the key to subjectivity, to experience.

I propose that the first and most elementary products of the protoself are "*primordial feelings*", which occur spontaneously and continuously whenever one is awake as well as in REM sleep dreaming. Such feelings provide a direct, wordless experience of one's own living body, thus connected to existence. They reflect the current state of the body along varied dimensions, for example, along the scale that ranges from pleasure to pain. They originate at the level of the brainstem rather than the cerebral cortex. Primordial feelings are a foundation for more complex levels of self. Feelings of emotion are variations on primordial feelings.

Pain and pleasure are body events mapped in a brain that at no instant is separated from its body. Primordial feelings are generated as a result of an ever-present body–brain interaction. The particular characteristics of the neural circuitry that accomplishes the connection, as well as certain properties of neurons, may also be relevant to the creation of feelings.

In this proposal, brains do not begin building conscious minds at the level of the cerebral cortex but rather at the level of the brainstem. Primordial feelings are not only the basic images generated by the brain but are also immediate manifestations of sentience, a view that runs counter to widely accepted thinking on brain and consciousness. On the other hand, the conscious mind as we know it also requires the cerebral cortices. They endow the mind-making process with the detailed images of the outside world that the brainstem cannot provide. In brief, conscious minds begin when brains add a self process to the on-going mind.

The self is built in distinct steps grounded on the protoself. The first step corresponds to primordial feelings, the elementary feelings of existence generated spontaneously from protoself activity. The second step is the core self. The core self is about a relationship between the organism and an object with which it interacts. The core self unfolds in a series of pulses, sequences of images that describe an object engaging the protoself and modifying that protoself, most notably its primordial feelings. The final step is the autobiographical self, which is defined in terms of biographical knowledge pertaining to the organism's past as well as to its anticipated future. The images that, in combination, define a biography generate pulses of core self whose aggregate constitutes an autobiographical self.

The protoself and its primordial feelings, along with the core self, constitute a basic self. The autobiographical self, whose higher reaches embrace all aspects of one's social persona, constitutes a robust personal and social self that can define identity. Normal human consciousness corresponds to a mind process in which all of these self levels operate, endowing a select number of mind contents with a momentary link to a pulse of core self.

At neither basic nor elaborate levels do self and consciousness happen in one area or region or center of the brain. Conscious minds result from the smoothly articulated operation of several brain systems. The ultimate consciousness products occur from numerous brain sites at the same time and not in one site in particular.

The key brain structures on which conscious minds depend include specific sectors of the upper brainstem, a set of nuclei in a region known as the thalamus, and specific but widespread regions of the cerebral cortex. These brain structures generate the lead triad of wakefulness, mind and self, which is found in standard conscious mind states (but not in dream consciousness, which is by definition an altered state of consciousness). All three structures contribute to some aspect of wakefulness, mind and self.

2.1 The Brainstem

The brainstem nuclei contribute to wakefulness, along with the hypothalamus. They are also responsible for constructing the protoself and generating primordial feelings. Significant aspects of the core self are implemented in the brainstem which also assists with the governance of attention. All these tasks require cooperation among the brainstem, the thalamus and the cerebral cortex.

A study of brainstem neuroanatomy reveals several distinct sectors of nuclei. A sector located at the bottom of the stem's vertical axis, largely in the medulla oblongata, contains the nuclei that are concerned with basic visceral regulation, notably cardiac and respiratory function. The destruction of these nuclei results in death. Above that level, in the pons and in the mesencephalon, there are nuclei whose damage has been associated with coma and vegetative state rather than death. This sector runs vertically from the mid-level of the pons to the top of the mesencephalon. It occupies the posterior part of the stem rather than the anterior, behind a vertical line that separates the anterior and posterior sectors of the brainstem. Two more structures relevant to consciousness are also located in the brainstem: the tectum and the hypothalamus. The tectum is the ensemble made by the superior and inferior colliculi. Besides their role in movements related to perception, the colliculi play a role in the coordination and integration of images. The hypothalamus is located immediately above the brainstem proper, but its deep involvement in life regulation and intricate interactions with brainstem nuclei justify its inclusion in the brainstem family.

The idea that certain sectors of the stem would be critical for consciousness but others would not, comes from classical neurological observations in cases of coma and vegetative state. The view presented here, however, differs in that it posits additional functional roles for the brain stem in the generation of consciousness.

Early in the history of consciousness research, the association between damage to the brainstem region and coma/vegetative state was taken to mean that the resulting dysfunction disrupted wakefulness or vigilance. The cerebral cortex was no longer energized and made active. Deprived of wakefulness, the mind was no longer conscious. The identification of a network of locally interactive neurons that projected upward, as a unit, toward the thalamus and cerebral cortex made this simple idea all the more plausible. The name given to this system of projections, the ascending reticular activating system or ARAS, captured the notion successfully. Such a system does exist and its projections are aimed at the intralaminar nuclei of the thalamus, which in turn project to the cerebral cortices. But the brainstem has other functions that relate to consciousness. In parallel with classical nuclei such as the cuneiform and pontis oralis, which are the origins for the ARAS, there are other nuclei, namely those involved in the management of internal body states, as well as others such as the locus coeruleus, the ventral tegmental nuclei and the raphe nuclei, that are responsible, respectively, for the release of norepinephrine, dopamine and serotonin in certain sectors of the cerebral cortex and basal forebrain. Most projections from these nuclei bypass the thalamus.

The main nuclei involved in body state management are the nucleus tractus solitarius (NTS) and the parabrachial nucleus (PBN), which I believe are involved in the creation of primordial feelings. The upper brainstem also includes the nuclei of the periaqueductal gray (PAG), whose activity results in behavioral and chemical responses that are part of life regulation and, in that capacity, also execute the emotions. The PAG nuclei are closely interlocked with those of the PBN and NTS and also with the deep layers of the superior colliculi, which are likely to play a coordinating role in the construction of the core self. The classical reticular nuclei and the ascending activating systems are indeed associated with wakefulness and sleep cycles, but the remaining brain stem nuclei participate in other important functions relevant to consciousness. These include (1) representation of standards for biological value, (2) representation of the organism's interior on the basis of which the protoself can be assembled and primordial feeling states generated, and (3) the first stages in the construction of the core self, which has consequences for the governance of attention.

These functions are related to the management of life, but in a broader sense they are not merely about in the regulation of viscera, metabolism and wakefulness. The brainstem nuclei are the neural home of the standard of biological value, and biological value has a pervasive influence throughout the brain in terms of structure and operations.

2.2 The Thalamus

Consciousness is often described as the result of massive integration of signals in the brain, and the thalamus is seen as playing a prominent role in that integration. But the thalamus plays a variety of functional roles pertinent to consciousness.

The thalamus contributes to all components of the standard conscious mind triad. Some thalamic nuclei are essential for wakefulness and to link brainstem to cortex; others bring in the inputs with which cortical maps can be assembled; still others assist with the integration that constitutes the background fabric of a complex mind.

The thalamus is a way-station for information collected from the body and destined to the cerebral cortex, including all the channels that carry signals about the body and about the world, such as pain and temperature, touch, hearing and vision; there is one exception, smell. All signals bound for the cerebral cortex stop over at thalamic relay nuclei.

The thalamus also processes the signals responsible for arousing the cerebral cortex or contributing to sleep, via neuron projections originating in the reticular formation, which change paths at the intralaminar nuclei on their way to structures such as the posteromedial cortices.

Most importantly, perhaps, from the standpoint of consciousness, the thalamus serves as a coordinator of cortical activities, a function that depends on the fact that several thalamic nuclei that signal to the cerebral cortex are in turn signaled back from the cerebral cortex, thus forming recursive loops. Such thalamic nuclei interconnect distant as well as close parts of the cerebral cortex. The purpose of the connectivity appears to be the interlinking of information.

The interplay between thalamus and cortex is likely to facilitate the simultaneous or sequenced activation of spatially separate patterns of activity, thus bringing them together in coherent patterns. Such activations would be responsible for the flow of images in one's stream of thought. This coordinating role is likely to depend on a back and forth between the associative thalamic nuclei and certain "choke" regions of the cerebral cortex, which are the regions that ensure convergence–divergence of signals and are also involved in coordinating cortical activities. In conclusion, the thalamus relays critical information to the cerebral cortex and inter-associates cortical information. The cerebral cortex and the thalamus co-evolved and neither can operate without the other.

2.3 The Cerebral Cortex

The human cerebral cortex is the current pinnacle of neural evolution. In interplay with thalamus and brainstem, the cerebral cortex maintains wakefulness and contributes to the selection of what we attend. In interplay with brainstem and thalamus, the cerebral cortex constructs the maps that become the mind. In interplay with the brainstem and the thalamus, the cerebral cortex helps generate the core self. Last, using the records of past activity stored in its memory banks, the cerebral cortex constructs our biography, complete with the experience of the physical and social environments we have inhabited. The cerebral cortex provides us with an identity and places that identity at the center of our conscious mind.

Assembling the consciousness show is such a cooperative effort that it would be unrealistic to single out any particular partner. We cannot engender the autobiographical aspects of self that define human consciousness without invoking the growth of convergence–divergence regions that dominate cortical neuroanatomy and neurophysiology (see Damasio 1989; Meyer and Damasio 2009). Autobiography could not arise without the contributions of the brainstem toward the protoself, or without the brainstem's linkage to the body proper, or the massive recursive integration engendered by the thalamus.

There is suggestive evidence that the early sensory cortices hold such explicit patterns. In typical multisensory experiences there are such simultaneous patterns, held in several separate regions whose ensemble can be appropriately designated as the brain's "image space". The imaged space is controlled functionally by the "dispositional space" (Meyer et al. 2010, 2011).

The systems vision outlined here dovetails with that of Stanislas Dehaene and Jean Pierre Changeux in their work on global neural workspace.

3 An Anatomical Bottleneck

There is a strange mismatch between the size of the human cerebral cortex and that of the human brainstem. Behind the functional alliance of these two key regions is an anatomical disproportion that only an evolutionary perspective can help explain.

Adjusted for body size, the basic design of the human brainstem dates back to reptilian times. The cerebral cortex of mammals, on the other hand, has expanded enormously in size and in architectural design, especially so in primates.

Because of its function as master life regulator, the brainstem has long been the recipient and local processor of the information required to represent the body and control its life. As it carried out this role in species whose cerebral cortex was minimal, the brainstem also developed the capacity to engender an elementary mind process and even a self process, via the protoself and core self mechanisms. The brainstem continues to carry out these same functions in humans. The greater complexity of the cerebral cortex, on the other hand, has allowed it to develop detailed image-making capabilities, imagination, expanded memory capacity and reasoning. It has also allowed it to develop language. The problem is that, notwithstanding the anatomical and functional expansion of the cerebral cortex, *the functions of the brainstem were never duplicated in the cortical structures*. The result is a convenient division of labor whose consequence is a complete interdependence of brainstem and cortex. Brainstem and cortex are forced into mutual cooperation.

This disproportion caused a major anatomo-functional bottleneck. Given that the brainstem was still being asked to ensure the full scope of life regulation and the foundations of consciousness, it was necessary to ensure that the brainstem influenced the cerebral cortex and that the activities of the cerebral cortex influenced the brainstem, most critically when it came to the construction of the core self. This is an especially important problem when we think that the detailed representation of most external objects and events only takes place in the cerebral cortex rather than in the brainstem.

The thalamus was the enabler of this accommodation. The thalamus accomplishes a dissemination of signals from the brainstem to a widespread territory of the cerebral cortex. In turn, the expanded cerebral cortex, directly and with the assistance of subcortical nuclei such as those in amygdalae and basal ganglia, funnels signals to the small-scale brainstem.

4 What Are Conscious Minds for?

Managing and safekeeping life is the fundamental premise of biological value. Biological value has influenced the evolution of brain structures and it influences numerous steps of brain operations. It is expressed as simply as in the release of chemical molecules related to reward and punishment or as elaborately as social emotions and sophisticated reasoning.

Consciousness plays a role in managing and safekeeping life efficiently. Neurological patients whose consciousness is disturbed are unable to manage their lives independently even when their basic life functions operate normally. However, mechanisms for managing and maintaining life are not novelties in biological evolution and did not depend on consciousness for their emergence. Such mechanisms already exist in single cells and are put in place and operated by the cells' genomes. They are present within un-minded and un-conscious neuron circuits, and they remain present in the phylogenetically older systems of human brains.

Conscious minds emerge within the history of life regulation. Life regulation, a dynamic process known as homeostasis, begins in unicellular living creatures, such as a bacterial cell or an amoeba, that do not have a brain but are capable of adaptive behavior. It progresses in individuals whose behavior is managed by simple brains, as is the case with worms, and it continues its march in individuals whose brains generate both behavior and mind. I suspect that whenever brains began to generate primordial feelings, and that is likely to be early in evolutionary history, organisms acquired a simple form of sentience. From there on, an organized self process developed providing the beginning of elaborate conscious minds. For example, reptiles qualify for this distinction, as do birds and mammals.

Most species whose brains generate a self do so at core level. As noted earlier, humans have both a core self and an autobiographical self. A number of mammals such as apes and cetaceans are also likely to have both.

The evolution of mind progress did not stop with the arrival of the self. Minds have become more complex, memory and reasoning have expanded notably, and the self processes have enlarged their scope. The core self has been gradually surrounded by an autobiographical self whose neural and mental specifications are different from those of the core self. We have become able to use a part of our mind's operation to monitor the operation of other parts. The conscious minds of humans, armed with complex selves and supported by strong capabilities of memory, reasoning, and language, created the instruments of culture and opened the way for new means of homeostasis at the level of societies and culture. Homeostasis was extended into the sociocultural space. Justice systems, economic and political organizations, the arts, medicine, and technology are examples of the new devices of regulation.

Basic homeostasis, which is non-consciously guided, and sociocultural homeostasis, which is created and guided by reflective conscious minds, are curators of biological value. Basic and sociocultural varieties of homeostasis are separated by billions of years of evolution, and yet they promote the same goal – the survival of living organisms – although in different ecological niches. Managing life in human brains requires both varieties of homeostasis, but the basic variety of homeostasis is an established set of devices, provided by everyone's genome, whereas the sociocultural variety is a work in progress.

The history of consciousness cannot be told in the conventional way. Consciousness contributed to more effective value management but it did not invent biological value or the process of valuation. Consciousness revealed biological value in human minds and allowed the development of new means of managing it.

Human consciousness and the functions that it made possible – language, expanded memory, reasoning, creativity – are the curators of value inside modern, minded, social beings.

References

Damasio AR (1989) Time-locked multiregional retroactivation: a systems level proposal for the neural substrates of recall and recognition. Cognition 33:25–62

Damasio A (2010) Self comes to mind. Pantheon, New York

Damasio A, Meyer K (2008) Consciousness: an overview of the phenomenon and of its possible neural basis. In: Laureys S, Tononi G (eds) The neurology of consciousness. Elsevier, London, pp 3–14

Meyer K, Damasio A (2009) Convergence and divergence in a neural architecture for recognition and memory. Trends Neurosci 32(7):376–382

Meyer K, Kaplan JT, Essex R, Webber C, Damasio H, Damasio A (2010) Predicting visual stimuli based on activity in auditory cortices. Nat Neurosci 13(6):667–668

Meyer K, Kaplan JT, Essex R, Damasio H, Damasio A (2011) Seeing touch is correlated with content-specific activity in primary somatosensory cortex. Cereb Cortex doi:10.1093/cercor/bhg289. First published online: February 17, 2011

The Global Neuronal Workspace Model of Conscious Access: From Neuronal Architectures to Clinical Applications

Stanislas Dehaene, Jean-Pierre Changeux, and Lionel Naccache

Abstract While a considerable body of experimental data has been accumulated on the differences between conscious and non-conscious processing, a theory is needed to bridge the neuro-psychological gap and establish a causal relationship between objective neurophysiological data and subjective reports. In the present review, we first briefly outline the detailed postulates and predictions of our working hypothesis, referred to as the *global neuronal workspace* (GNW) model. We then compare these predictions to experimental studies that have attempted to delineate the physiological signatures of conscious sensory perception by contrasting it with subliminal processing, using a variety of methods: behavioral, PET and fMRI imaging, time-resolved imaging with ERP and MEG, and finally single-cell electrophysiology. In a final section, we examine the relevance of these findings for pathologies of consciousness in coma and vegetative states.

1 Introduction

In recent years, experimental methods have become available to investigate conscious and non-conscious processing, and a considerable body of experimental data has accumulated aiming to establish a causal relationship between objective and subjective data on conscious access. In parallel, computational neuroscience has developed as a fast-moving discipline, with the aim of building up formal mathematical or simulation models, expressed in terms of neuronal networks, that link

S. Dehaene (✉)
INSERM, Cognitive Neuroimaging Unit, Gif sur Yvette 91191, France
and
CEA, DSV, I2BM, Neurospin center, Gif sur Yvette 91191, France
and
University Paris 11, Orsay, France
and
Collège de France, 11 Place Marcelin Berthelot, 75005, Paris, France
e-mail: Stanislas.Dehaene@cea.fr

S. Dehaene and Y. Christen (eds.), *Characterizing Consciousness: From Cognition to the Clinic?* Research and Perspectives in Neurosciences,
DOI 10.1007/978-3-642-18015-6_4, © Springer-Verlag Berlin Heidelberg 2011

together the molecular, neuronal, physiological and behavioral/mental data in a coherent, non-contradictory though minimal form (Changeux and Dehaene 1989). Implemented as "formal organisms", these neuro-computational models should ultimately address the challenge of decades of attempts to account for subjective experience, which provides a unified or global mental scene where a synthesis between past, present and future takes place and where multimodal perceptions, emotions and feelings (present), evoked memories of prior experiences (past), together with anticipations of actions (future) become subjectively integrated in a continuously changing and dynamic stream of consciousness (Baars 1989; Crick and Koch 2003; Dehaene et al. 1998; Dehaene and Naccache 2001; Edelman 1989; James 1890).

In the present review, we first briefly outline the detailed postulates and predictions of our working hypothesis, referred to as the global neuronal workspace (GNW) model. We then examine the experimental studies that have attempted to delineate the physiological signatures of conscious sensory perception by contrasting it with subliminal processing, using a variety of methods: behavioral, PET and fMRI imaging, time-resolved imaging with ERP and MEG, and finally single-cell electrophysiology. In a final section, we examine the relevance of these findings for pathologies of consciousness in coma and vegetative states.

2 A Brief Overview of the GNW Model

The GNW model in its original formulation (Dehaene and Changeux 2005; Dehaene et al. 1998, 2003, 2006; Dehaene and Naccache 2001) relies upon a few simple assumptions. Its main postulate is that conscious access is global information availability (see Baars 1989): what we subjectively experience as conscious access is the selection, amplification and global broadcasting, to many distant areas, of a single piece of information selected for its salience or relevance to current goals.

From a neuronal architecture standpoint, two main computational spaces are distinguished within the brain, each characterized by a distinct pattern of connectivity (see Fig. 1): (1) a processing network, composed of a set of parallel, distributed and functionally specialized processors (Baars 1989) or modular subsystems (Shallice 1988) subsumed by topologically distinct cortical domains with highly specific local or medium-range connections that "encapsulate" information relevant to its function. Processors typically operate non-consciously and in a bottom-up manner, although local top-down projections may also contribute to their operation by providing local predictions and prediction errors (see e.g. Bekinschtein et al. 2009); and (2) a global neuronal workspace (GNW), consisting of a distributed set of cortical neurons characterized by their ability to receive from and send back to homologous neurons in other cortical areas horizontal projections through long-range excitatory axons. Such long-range cortico-cortical connections

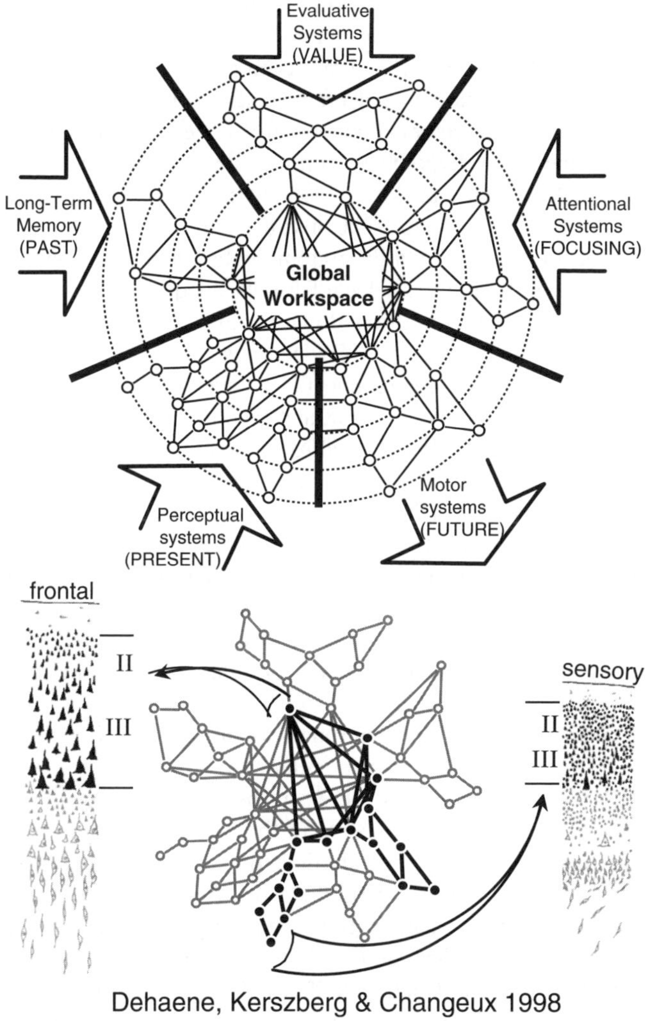

Fig. 1 Schema of the global neuronal workspace (GNW) model. The GNW model proposes that associative perceptual, motor, attention, memory and value areas interconnect to form a higher-level unified space where information is broadly shared and broadcasted back to lower-level processors. The GNW is characterized by its massive connectivity, made possible by thick layers II/III with large pyramidal cells sending long-distance cortico-cortical axons, particularly dense in PFC (after Dehaene et al. 1998)

include callosal connections and mostly originate from the pyramidal cells of layers 2 and 3 that are particularly elevated in prefrontal, parieto-temporal and cingulate associative cortices, together with their thalamo-cortical relationships (Von Economo 1929). GNW neurons typically accumulate information through recurrent top–down/bottom–up loops, in a competitive manner such that a single representation eventually achieves a global conscious status. Because GNW neurons are broadly distributed, there is no single brain center where conscious information is gathered and dispatched but rather a brain-scale process of conscious synthesis achieved when multiple processors converge to a coherent metastable state.

According to the GNW hypothesis, conscious access proceeds in two successive phases (see also Chun and Potter 1995; Lamme and Roelfsema 2000). In a first phase, lasting from ~100 to ~300 ms, the stimulus climbs up the cortical hierarchy of processors in a primarily bottom–up and non-conscious manner. In a second phase, if the stimulus is selected for its adequacy to current goals and attention state, it is amplified in a top–down manner and becomes maintained by sustained activity of a fraction of GNW neurons, the rest being inhibited. The entire workspace is globally interconnected in such a way that only one such conscious representation can be active at any given time (see Sergent et al. 2005; Sigman and Dehaene 2005, 2008). This all-or-none invasive property distinguishes it from peripheral processors in which, due to local patterns of connections, several representations with different formats may coexist. Simulations, further detailed below, indicate that the late global phase is characterized by several unique features. These predicted "signatures" of conscious access include a sudden, late and sustained firing in GNW neurons [referred to as conscious "ignition" in Dehaene et al. (2003)], a late sensory amplification in relevant processor neurons, and an increase in high-frequency oscillations and long-distance phase synchrony.

The state of activation of GNW neurons is assumed to be globally regulated by vigilance signals from the ascending reticular activating system that are powerful enough to control major transitions between the awake state (GNW active) and slow-wave sleep (GNW inactive) states. In the resting awake state, the brain is the seat of an important ongoing metabolic activity (Gusnard and Raichle 2001). An important statement of the GNW model is that the GNW network is the seat of a particular kind of brain-scale activity state characterized by spontaneous "ignitions" similar to those that can be elicited by external stimuli, but occurring endogenously (Dehaene and Changeux 2005). A representation that has invaded the workspace may remain active in an autonomous manner and resist changes in peripheral activity. If it is negatively evaluated, or if attention fails, it may, however, be spontaneously and randomly replaced by another discrete combination of workspace neurons, thus implementing an active "generator of diversity" that constantly projects and tests hypotheses on the outside world (Dehaene and Changeux 1989, 1991, 1997). The dynamics of workspace neuron activity is thus characterized by a constant flow of individual coherent episodes of variable duration, selected by specialized reward processors.

Several explicit formulations and computer simulations of the GNW architecture and physiology were successively developed (Dehaene and Changeux 2005; Dehaene et al. 1998, 2003, 2006; Zylberberg et al. 2009), leading to specific experimental predictions.

3 Simulations of the Neural Signatures of Conscious Access

The GNW has been implemented as explicit computer simulations of neural networks (Dehaene and Changeux 2005; Dehaene et al. 1998, 2003; see also Zylberberg et al. 2009). These simulations incorporate spiking neurons and synapses with detailed membrane, ion channel and receptor properties organized into distinct cortical supra-granular, granular, infra-granular and thalamic sectors with reasonable connectivity and temporal delays. Although the full GNW architecture was not simulated, four areas were selected and hierarchically interconnected (Fig. 2). Bottom–up feed-forward connections linked each area to the next whereas long-distance top–down connections projected to all preceding areas. Moreover, in a simplifying assumption supported by recent data (Self et al. 2008), bottom–up connections impinged on

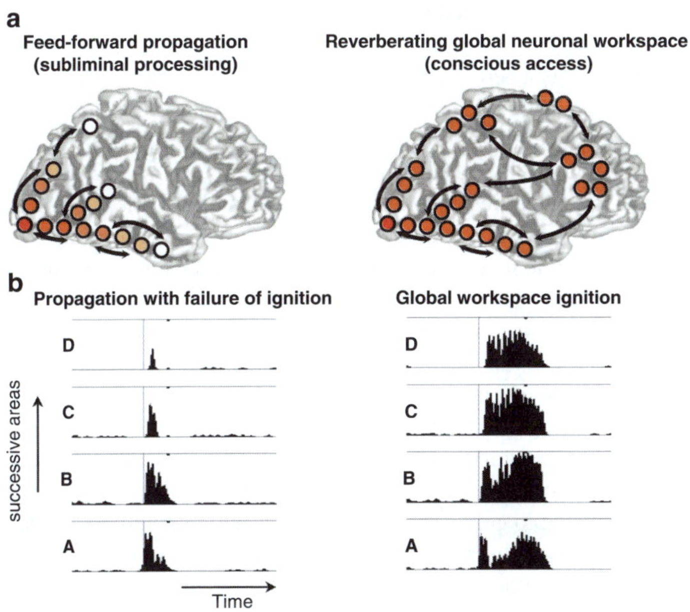

Fig. 2 Events leading to conscious access according to the GNW model. (**a**) Main postulated differences between subliminal and conscious processing. During feed-forward propagation, sensory inputs progress through a hierarchy of sensory areas in a feed-forward manner, successively contacting diverse and not necessarily compatible representations corresponding to all probabilistic interpretations of the stimuli. Multiple signals converge to support each other's interpretation in higher-level cortical areas. Higher areas feedback onto lower-level sensory representations, favoring a convergence towards a single coherent representation compatible with current goals. (**b**) Simulation of two single trials in which an identical pulse of brief stimulation was applied to sensory inputs (Dehaene and Changeux 2005). Fluctuations in ongoing activity prevented ignition in the left diagram, resulting in a purely feedforward propagation dying out in higher-level areas. In the right diagram, the same stimulus crossed the threshold for ignition, resulting in self-amplification, a global state of activation, oscillation and synchrony, and a late long-lasting wave of late activation reaching back to early sensory areas

glutamate AMPA receptors whereas the top–down ones, which are slower, more numerous and more diffuse, primarily involved glutamate NMDA receptors (yet does not include, at this stage, the functionally and pharmacologically important set of metabotropic glutamate receptors; see Niswender and Conn 2010). In higher areas, inputs competed with each other through GABAergic inhibitory interneurons, and it was assumed (though not explicitly simulated) that the winning representation would be broadcasted by additional long-distance connections to yet other cortical regions.

Our simulations first explored the sequence of activity leading to conscious access. When sensory stimulation was simulated, using a brief depolarizing input at the lowest thalamic level, activation propagated according to two successive phases (see Fig. 2): (1) initially, a brief wave of excitation progressed into the simulated hierarchy through fast AMPA-mediated feedforward connections, with an amplitude and duration directly related to the initial input; and (2) in a second stage, mediated by the slower NMDA-mediated feedback connections, the advancing feed-forward wave amplified its own inputs in a cascading manner, quickly leading the whole network into a global self-sustained reverberating or "ignited" state. This ignition was characterized by an increased power of local cortico-thalamic oscillations in the gamma band (Dehaene et al. 2003), as well as increased synchrony across distant regions (Dehaene, unpublished simulations). Thus, this second phase of the simulation reproduces the signatures of conscious access detailed in the empirical section – late, all-or-none, cortically distributed potentials involving prefrontal cortex (PFC) and other high-level associative cortices, with simultaneous increases in high-frequency power and synchrony (e.g. Gaillard et al. 2009). GNW simulations capture both the initial linear response and the late perception threshold present in experimental data (de Lafuente and Romo 2006; Del Cul et al. 2007).

Further simulations indicated that ignition could fail to be triggered under specific conditions, thus leading to simulated non-conscious states. For very brief or low-amplitude stimuli, a feed-forward wave was seen in the initial thalamic and cortical stages of the simulation, but it died out without triggering the late global activation because it was not able to gather sufficient, self-sustaining reverberant (Dehaene and Changeux 2005). Even at higher stimulus amplitudes, the second global phase could also be disrupted if another incoming stimulus had been simultaneously accessed (Dehaene et al. 2003). Such a disruption occurs because, during ignition, the GNW is mobilized as a whole, some GNW neurons being active while the rest are actively inhibited, thus preventing multiple simultaneous ignitions. A strict seriality of conscious access and processing is therefore predicted and has been simulated (Dehaene and Changeux 2005; Dehaene et al. 2003; Zylberberg et al. 2010). Overall, these simulations thus captured the two main types of experimental conditions known to lead to non-conscious processing: stimulus degradation (e.g., masking) and inattention due to a simultaneous task (e.g., attentional blink).

The transition to the ignited state could be described, in theoretical physics terms, as a stochastic phase transition – a sudden change in neuronal dynamics whose occurrence depended in part on stimulus characteristics and in part on spontaneous fluctuations in activity (Dehaene and Changeux 2005; Dehaene et al. 2003). Pre-stimulus fluctuations in neural discharges only had a small effect on the early sensory stage, which largely reflected objective stimulus amplitude and duration, but they had a large influence on the second stage, which was characterized by a bimodal, "all-or-none" distribution of activity, similar to empirical observations (Quiroga et al. 2008; Sergent et al. 2005; Sergent and Dehaene 2004). Across trials, the very same stimulus did or did not lead to global ignition, depending on the precise phase of the stimulus relative to ongoing spontaneous activity. This notion that pre-stimulus baseline fluctuations partially predict conscious perception is now backed up by considerable empirical data (e.g., Boly et al. 2007; Palva et al. 2005; Sadaghiani et al. 2009; Super et al. 2003; Wyart and Tallon-Baudry 2009).

In GNW simulations, conscious broadcasting manifests itself as a slow depolarization of layer II/III apical dendrites of pyramidal dendrites in a subset of target GNW neurons. The many post-synaptic potentials evoked by this long-distance, cortico-cortical signalling would therefore create slow intracellular currents travelling from the apical dendrites towards the cell's soma, summing up on the cortical surface as negative slow cortical potentials (SCPs), which have indeed been suggested to be associated with conscious processing (for review, see He and Raichle 2009). In scalp EEG measurements, negative potentials indeed predominate over regions dense in active GNW neurons, and their topography reflects which content is currently active in working memory (e.g., Vogel and Machizawa 2004). However, ignition does not imply that all GNW neurons are simultaneously active. In the simulations, a subset of neurons is activated while the rest is strongly suppressed by lateral inhibition via GABAergic interneurons, thus leading to attentional blink and serial processing (Dehaene et al. 2003). Which neurons are active and which are inactive defines the current conscious content. The impact of this broad GNW inhibition may perhaps be easier to detect at the scalp level, particularly if a sparse code is used, where the set of active neurons is relatively small, e.g., 10% (for neurophysiological evidence, see e.g., Deco et al. 2007). Inhibition, by the same logic as above, should result in broad *positive* surface potentials. As already noted by Rockstroh et al. (1992, p. 175), "the surface positivity corresponding to these inhibited networks would then dominate over the relatively smaller spots of negativity caused by the reverberating excitation." Thus, the GNW hypothesis, combined with elementary physiological arguments, can explain why the P3b component (a broadly distributed scalp positivity) is a frequent neurophysiological signature of conscious access and can draw attention to the important putative role of GABAergic interneurons and their neuromodulatory control in conscious processing.

4 Simulations of the Impact of Vigilance on Conscious Access

Further simulations explored the impact of vigilance on access to the GNW (Dehaene and Changeux 2005). Changes in vigilance, including the transition between the awake and asleep states, are regulated by various diffuse ascending neuromodulatory systems located in the brainstem, hypothalamus, and basal forebrain and by liberating substances such as acetylcholine, noradrenalin, serotonin, and histamine in the cortex and thalamus. For simplicity, we summarized these complex effects of ascending neuromodulatory inputs by a change in a single parameter, the amount of depolarizing current injected into both thalamic and cortical pyramidal neurons (Dehaene and Changeux 2005).

The model exhibited two interesting changes when the neuromodulatory current was increased. First, there was a threshold value of ascending neuromodulation below which very little spontaneous activity was seen but above which structured neuronal activity emerged in the form of spontaneous thalamocortical oscillations in the gamma band. This GNW prediction fits with the sudden emergence of a broad network of GNW areas, including PFC and its thalamic connections, during awakening or recovery from vegetative state (Balkin et al. 2002; Laureys et al. 2000), and its drastic disruption during anesthesia, coma, or slow wave sleep (reviewed in Laureys 2005, and further discussed below).

Second, when a fixed external sensory stimulus was presented, the threshold for global ignition of the GNW network decreased as the strength of the ascending neuromodulation signal increased (Dehaene and Changeux 2005). At one extreme, very low levels of ascending inputs prevented any ignition, even by long and intense stimuli: in this case, only a small transient activation was seen in the lower levels of the model. These simulations can be compared with empirical observations of sensory processing during the vegetative state, where activation can be seen in the thalamus and early sensory cortices, but fail to extend to the prefrontal, parietal, and cingulate regions seen in awake normal subjects (see below).

5 Simulations of Spontaneous Activity and Serial Processing

One last important feature of the GNW model is the occurrence of highly structured spontaneous activity. Even in the absence of external inputs, the simulated GNW neurons are assumed to fire spontaneously, in a top–down manner, starting from the highest hierarchical levels of the simulation and propagating downwards to form globally synchronized ignited states (Dehaene and Changeux 2005). When the ascending vigilance signal is large, several such spontaneous ignitions follow each other in a never-ending stream and partially prevent access to external stimuli, possibly capturing empirical observations of inattentional blindness (Mack and Rock 1998) and mind wandering (Christoff et al. 2009; Mason et al. 2007; Smallwood et al. 2008).

In initial simulations, a spike-rate adaptation current was solely responsible for the interruption of an ignited state and its replacement by another. However, more complex network architectures have also been simulated in which a goal state is set and where structured patterns of activity are spontaneously generated until the goal is attained (Dehaene and Changeux 1997; Zylberberg et al., 2010). In these simulations, ignited states are only metastable and can be destabilized by a negative reward that signals deviation from the current goal, in which case they are spontaneously and randomly replaced by another discrete combination of workspace neurons, thus implementing an active generator of diversity that constantly projects and tests hypotheses on the outside world (Dehaene and Changeux 1989, 1991, 1997). The dynamics of such networks is characterized by a constant flow of individual coherent episodes of variable duration, selected by reward signals to achieve a defined goal state. Architectures based on these notions have been applied to a variety of tasks (delayed response: Dehaene and Changeux 1989; Wisconsin card sorting: Dehaene and Changeux 1991; Tower of London: Dehaene and Changeux 1997; Stroop: Dehaene et al. 1998), although a single architecture common to all tasks is not yet in sight (but see Rougier et al. 2005). They provide a preliminary account of why GNW networks are spontaneously active, in a sustained manner, during effortful tasks that require series of conscious operations, including search, dual-task and error processing.

In summary, simulations of the GNW architecture explain the close similarity of the brain activations seen during (1) conscious access to a single external stimulus, (2) modulation of conscious access by changes in vigilance and (3) effortful serial processing, even in the absence of an external stimulus. In all of these cases, the GNW provides a working memory space that can be temporarily detached from incoming stimuli and operates in an autonomous manner.

6 Anatomy of the GNW

The main anatomical premise of the GNW model is that it consists of "a distributed set of cortical neurons characterized by their ability to receive from and send back to homologous neurons in other cortical areas horizontal projections through long-range excitatory axons mostly originating from the pyramidal cells of layers 2 and 3" (Dehaene et al. 1998). It is grounded in the "special morphology" of the pyramidal cells from the cerebral cortex already noted by Cajal et al. (1892/1988), who mentioned their "long axons with multiple collaterals" and their "very numerous and complex dendrites", and also noted that these fibers of association form "the main mass of the white matter." Recent investigations have confirmed the view that the cortico-cortical and callosal fibers primarily (though not exclusively) arise from layer II–III pyramids. Von Economo (1929) further noted that the cerebral cortex is not uniform in structure and that these large pyramidal cells in layers III and V are specially abundant in areas "spread over the anterior two-thirds of the frontal lobe, (...) the superior parietal lobule" and "the cingulate cortex", among other cortical areas.

Pyramidal cells comprise over 70% of all neurons in the cerebral cortex (DeFelipe and Farinas 1992) and display a remarkable diversity. Quantitative analyses of the dendritic field morphology of layer III pyramidal neurons revealed a continuous increase of complexity of the basal dendrites from the occipital up to the PFC within a given species (DeFelipe and Farinas 1992; Elston and Rosa 1997,

1998) and from lower species (owl monkey, marmoset) up to humans (Elston 2003). Consistent with its proposed role in the GNW, the PFC receives the most diverse set of cortico-cortical inputs from a multitude of areas involved in processing all sensory modalities (Cavada et al. 2000; Fuster 2008; Kringelbach and Rolls 2004; Pandya and Yeterian 1990; Petrides and Pandya 2009). In particular, mapping of long-range connections in the monkey cerebral cortex revealed long-range connections linking, among others, the PFC (area 46), the superior temporal sulcus, parietal area 7a and the hippocampus, together with the contralateral anterior and posterior cingulum, area 19 and the parahippocampal gyrus (Goldman-Rakic 1988). These circuits were suggested to contribute to working memory and the central executive (Goldman-Rakic 1996). In addition, areas within PFC are multiply interconnected (Barbas and Pandya 1989; Preuss and Goldman-Rakic 1991), and the superficial layers in PFC are characterized by the abundance of horizontal intrinsic axon projections that arise from supragranular pyramidal cells (Kritzer and Goldman-Rakic 1995; Melchitzky et al. 1998, 2001; Pucak et al. 1996), thus exhibiting the massive and recurrent interconnectivity needed to sustain workspace states.

In the course of postnatal brain development, dendritic and axonal arborizations develop progressively through nested processes of neurite outgrowth and selection through synapse selective stabilization and elimination or "pruning" (Bourgeois et al. 1994; Changeux et al. 1973; Changeux and Danchin 1976). The data are consistent with the view that most (about 2/3 in the monkey) of the cortical synaptic endowment of the GNW develops postnatally, followed by a consistent and continuous maturation manifested by a decrease of the global synaptic envelope down to the adult brain synaptic content.

Identification of the detailed anatomy of white matter pathways interconnecting cortical areas started with Theodore Meynert (1833–1892), who classified white matter fibers into association fibers connecting cortical areas within a hemisphere, commissural fibers connecting the two hemispheres, and projection fibers, ascending or descending pathways arising and terminating in the cortex. In a still up-to-date volume, Dejerine (1895) further distinguished U-shaped, short, *intra*-area association fibers that remain superficial close to the gray matter and long *inter*-areas association fibers that run more deeply in the white matter. Among this last category Dejerine distinguished five main tracts: the cingulum that connects medial frontal, parietal, occipital, temporal and cingulate cortices, the uncinate fasciculus that connects orbitofrontal to anterior and medial temporal lobes, the arcuate/superior longitudinal fasciculus that connects perisylvian frontal, parietal and temporal cortices, the inferior fronto-occipital fasciculus that connects the orbital and lateral frontal cortices to occipital cortex and the inferior longitudinal fasciculus that connects the occipital and temporal lobes. Consistent with the GNW hypothesis, four of them connect the frontal areas with other cortical areas.

The course of cortical tracts can now be confirmed by diffusion tensor imaging and tractography algorithms, yet with important limitations. Measurements typically average over relatively large voxels (a few millimeters aside) that contain a diversity of criss-crossing fibers. Current tracking algorithms suffer from severe difficulties in estimating the true course of fibers, especially when a relatively smaller tract crosses a large one. Furthermore, long-distance fibers, are typically underestimated by probabilistic tracking algorithms, unless distance compensation is introduced. Even recent articles claiming to study the entire "connectome" (e.g., Hagmann et al. 2008) suffer from a massive underestimation of the true longdistance connectivity of areas 46, 6, FEF (frontal eye fields) and LIP (lateral intraparietal sulcus), critical to GNW theory and known from macaque invasive tracer studies, careful human anatomical dissections dating from the end of the nineteenth century (Dejerine, Meynert, Flechsig), and some reliable diffusion tensor tractography methods (Catani and Thiebaut de Schotten 2008).

Because of these limitations, it is important to confirm the inferred networks using converging techniques. The correlation of cortical thickness measures at two different cortical sites across many individuals can serve as a proxy for cortico-cortical connectivity (Bassett et al. 2008; He et al. 2009). There is a remarkable convergence of this method with fMRI intercorrelation during the resting state or by phase synchrony in the beta band during either working memory (Bassett et al. 2009) or conscious access in the attentional blink (Gross et al. 2004).

Human structural and functional brain networks have been further explored using graph theory (Bassett and Bullmore 2009; Bullmore and Sporns 2009; Gong et al. 2009; Hagmann et al. 2008; Iturria-Medina et al. 2008). Here brain networks are described as composed of nodes (brain regions) linked by edges (axonal projections). The resulting undirected graph can then be characterized by parameters such as degree (number of connections from a given node), the minimal path length between any two nodes, and clustering (density of triangular connections). Graph analyses indicate that the brain lives in the small-world regime (Watts and Strogatz 1998) characterized by a small path length, indicating a high transmission efficiency, and a high level of clustering, providing a tolerance of the overall system to the failure of individual nodes or connections. Putative hubs (nodes with a high degree of connectivity) were identified within the posterior medial and parietal cerebral cortex, including the precuneus, the insula, the superior parietal cortex and to some extent the superior frontal cortex (though these results are probably limited by the biases noted above; Gong et al. 2009; Hagmann et al. 2008; Iturria-Medina et al. 2008). Particularly notable is the prominent structural role of the precuneus, a region homologous to the highly connected posteromedial cortex (PMC) in the macaque (Parvizi et al. 2006). Being an aggregate of convergence-divergence zones (Meyer and Damasio 2009) and tightly connected to PFC area 46 and other workspace regions (Goldman-Rakic 1999), the PMC may play a critical role in humans in the "generation of the self process" (Damasio 1999) and in self-referential processing (Cavanna and Trimble 2006; Vogt and Laureys 2005). As discussed below, deactivation of PMC is seen during loss of consciousness in coma and vegetative state, anesthesia and slow wave sleep (Alkire et al. 2008; Laureys 2005). In a general manner, these results are thus consistent with the GNW model assumption of a tight relation between anatomical long-distance association fibers and functional conscious access.

7 Imaging States of Conscious Access and Non-Conscious Processing

As noted by Baars (1989), the experimental study of the mechanisms of conscious access requires the definition of a minimal contrast between a situation in which information is consciously accessed and a similar situation in which the same information is only processed non-consciously. Many such contrasts are now available (Kim and Blake 2005). Our own brain imaging work relied primarily on two techniques: retrograde masking, where the stimulus is flashed for a perceptible duration but is made invisible by the subsequent presentation, at the same location, of another shape, called the "mask;" and the attentional blink (AB), where a brief target, presented for a duration that would be perceivable in isolation, becomes invisible once the participants are temporarily distracted by a concurrent task. In both cases, functional magnetic resonance imaging (fMRI), magneto-encephalography (MEG), electro-encephalography (EEG) and intracranial recordings can be used to record the progression of activation in the cortical hierarchy under conditions of conscious versus non-conscious perception.

Such a research program requires a consensus on an empirical criterion to discriminate conscious and non-conscious processing. According to a long psychophysical tradition, grounded in signal-detection theory, a stimulus should be accepted as non-conscious or "subliminal" (below threshold) only if subjects are unable to perform above chance on some direct task of stimulus detection or classification. This objective definition raises problems, however (Persaud et al. 2007; Schurger and Sher 2008). First, it tends to overestimate conscious perception: there are many conditions in which subjects perform better than chance, yet still deny perceiving the stimulus. Second, performance can be at chance level for some tasks but not others, raising the issue of which tasks count as evidence of conscious perception or merely of subliminal processing. Third, the approach requires accepting the null hypothesis of chance-level performance, yet performance never really falls down to zero, and whether it is significant or not often depends on arbitrary choices such as the number of trials dedicated to its measurement. For these reasons, our research has emphasized obtaining subjective reports of stimulus visibility, if possible on every single trial (Sergent and Dehaene 2004). Such subjective reports are arguably the primary data of interest in consciousness research. Furthermore, reports of stimulus visibility can be finely quantified, leading to the discovery that conscious perception can be "all-or-none" in some masking and AB paradigms (Del Cul et al. 2006, 2007; Sergent and Dehaene 2004). Subjective reports also present the advantage of assessing conscious access immediately and on every trial, thus permitting post-experiment sorting of conscious versus non-conscious trials with identical stimuli (e.g., Del Cul et al. 2007; Lamy et al. 2009; Pins and Ffytche 2003; Sergent et al. 2005; Wyart and Tallon-Baudry 2008). Importantly, objective assessments, wagering indices and subjective reports are generally in excellent agreement (Del Cul et al. 2006, 2009; Persaud et al. 2007). For instance, the masking thresholds derived from objective and subjective data are essentially identical across subjects ($r = 0.96$, slope ≈ 1) (Del Cul et al. 2006).

Those data suggest that conscious access causes a major change in the availability of information that is easily detected by a variety of subjective and objective measures.

Once criteria for conscious access are defined, one can examine how the brain activity evoked by accessed stimuli differs from that elicited by well-matched stimuli that remain subjectively invisible. Here, we successively review the four levels of analysis at which such experiments have been attempted: fMRI, time-resolved methods such as event-related potentials (ERPs) and MEG, intracranial recordings, and finally, single-cell recordings.

7.1 Brain-Scale Neuroimaging with fMRI

fMRI can provide a global image of the brain activity evoked by a visible or invisible stimulus integrated over a few seconds. Grill-Spector et al. (2000) first used fMRI to measure visual activity evoked by masked pictures presented below or above the visibility threshold. V1 activation was largely unaffected by masking, but the amount of activation in more anterior regions of lateral occipital and fusiform cortex strongly correlated with perceptual reports. A year later (Dehaene et al. 2001), a similar contrast between masked and unmasked words, now at the whole-brain level, again revealed a strong correlation of conscious perception with fusiform activity but also demonstrated extended areas of activation uniquely evoked by conscious words, including inferior prefrontal, mesial frontal and parietal sites. In more recent fMRI work, under conditions where masking followed a characteristic U-shaped curve, fusiform and midline prefrontal and inferior parietal regions again closely tracked conscious perception (Haynes et al. 2005). An important control was recently added: participants' objective performance could be equated while subjective visibility was manipulated (Lau and Passingham 2006). In this case, a correlate of visibility could only be detected in left dorsolateral PFC.

Some authors have found correlations of fMRI activation with visibility of masked versus unmasked stimuli exclusively in posterior visual areas (e.g., Tse et al. 2005). However, in their paradigm, even the unmasked stimuli were probably not seen because they were unattended and irrelevant, which can prevent conscious access (Dehaene et al. 2006; Kouider et al. 2007; Mack and Rock 1998). Overall, fMRI evidence suggests two convergent correlates of conscious access: (1) amplification of activity in visual cortex, clearest in higher-visual areas such as the fusiform gyrus, but sometimes including area V1 (Williams et al. 2008); and (2) emergence of a distributed set of areas, virtually always including bilateral parietal and prefrontal cortices (see Fig. 3).

7.2 Time-Resolved Imaging Methods

ERPs and MEG have been used to track the processing of a stimulus in time, at a millisecond scale, as it crosses or does not cross the threshold for subjective

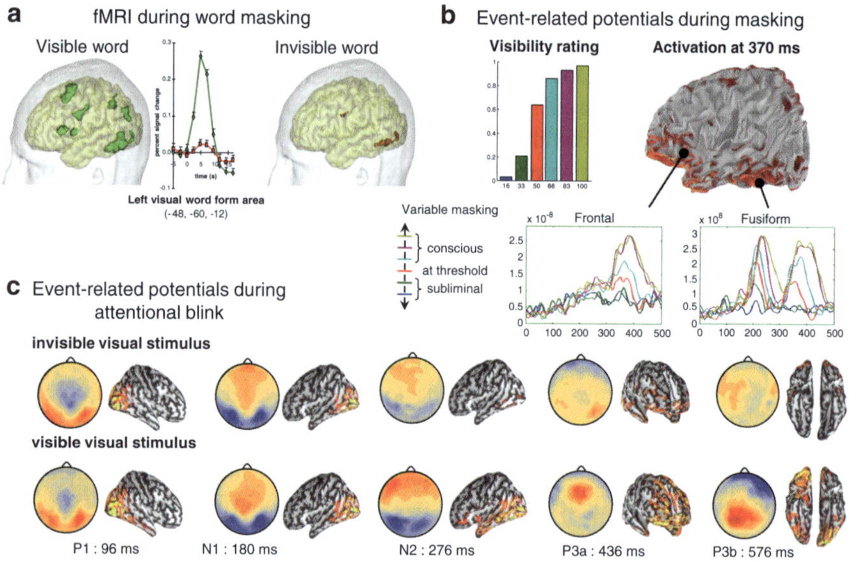

Fig. 3 Brain-imaging correlates of conscious access. (**a**) An early fMRI experiment contrasting the fMRI activations evoked by brief presentations of words that were either normally readable (*left*) or made invisible by masking (*right*) (Dehaene et al. 2001). Non-conscious word processing activated the left occipito-temporal visual word form area, but conscious perception was characterized by (1) an intense amplification of activation in relevant non-conscious processors, here the visual word form area (*left* occipito-temporal cortex; see *middle* graph); and (2) an additional spread of activation to a distributed, though restricted set of associative cortices including inferior parietal, prefrontal and cingulate areas. (**b**) Event-related potentials during masking, with manipulation of visibility by varying the temporal asynchrony between a visual stimulus and a subsequent mask (Del Cul et al. 2007). A non-linearity, defining a threshold value for conscious access, was seen in both subjective visibility reports and the P3b event amplitude. Source modelling related this P3b to a sudden non-linear ignition, about 300 ms after stimulus presentation, of distributed sources including inferior PFC, with a simultaneous re-activation of early visual areas. Note the two-stage pattern of fusiform activation, with an early linear activation followed by a late non-linear ignition. (**c**) Time course of scalp event-related potentials evoked by an identical visual stimulus, presented during the attentional blink, as a function of whether it was reported as seen or unseen (Sergent et al. 2005). Early events (P1 and N1) were strictly identical, but the N2 event was amplified and the P3 events (P3a and P3b) were present essentially only during conscious perception

report. Early visual activation can be fully preserved during masking (Schiller and Chorover 1966), a finding supported by both animal electrophysiology (Bridgeman 1975, 1988; Kovacs et al. 1995; Lamme et al. 2002; Rolls et al. 1999) and essentially all recent ERP and MEG studies (Dehaene et al. 2001; Del Cul et al. 2007; Fahrenfort et al. 2007; Koivisto et al. 2006, 2009; Lamy et al. 2009; Melloni et al. 2007; Railo and Koivisto 2009; van Aalderen-Smeets et al. 2006; for an exception, see Pins and Ffytche 2003). Evidence from the AB confirms that the first 200 ms of initial visual processing can be fully preserved on trials in which subjects deny seeing a stimulus (Sergent et al. 2005; Vogel et al. 1998; see Fig. 3).

When does a consistent correlation with subjective reports of visibility first emerge? Following the initial P1 (~100 ms) and N1 (~170 ms) waves, at around 200–300 ms, more negative voltages are seen over posterior cortices under conditions that are conducive to higher visibility (Del Cul et al. 2007; Fahrenfort et al. 2007; Koivisto et al. 2008, 2009; Railo and Koivisto 2009; Sergent et al. 2005). Koivisto and collaborators have called this event the visual awareness negativity (VAN). Lamme and collaborators (Fahrenfort et al. 2007) find an even earlier amplification in visual cortex, just posterior to the P1 wave (110–140 ms). Several arguments, however, mitigate the possibility that these mid-latency differences already reflect conscious perception. First, they are absent from several carefully controlled experiments (e.g., Lamy et al. 2009; van Aalderen-Smeets et al. 2006). Second, their profile of variation with stimulus variables such as target-mask delay does not quite mimic the subject's conscious reports (Del Cul et al. 2007; van Aalderen-Smeets et al. 2006). Indeed, they typically consist in differences in degree, modulating N1 and P2 events that remain strongly present on non-conscious trials (Del Cul et al. 2007; Fahrenfort et al. 2007; Sergent et al. 2005). Although more research is needed to clarify this point, they might represent a cascading modulation of non-conscious processing, perhaps causally related to subsequent conscious access but not yet reflective of the all-or-none process of conscious perception itself.

The most consistent correlate of visibility appears to be a late (~300–500 ms), broadly distributed, positive component called P3 or sometimes P3b (to distinguish it from the focal anterior P3a, which is thought to reflect automatic attention attraction and can occur non-consciously; see, e.g., Muller-Gass et al. 2007; Salisbury et al. 1992). This component is reproducibly observed as strongly correlated with subjective reports, both when varying stimulus parameters (e.g. Del Cul et al. 2007) and when comparing identical trials with or without conscious perception (e.g., Babiloni et al. 2006; Del Cul et al. 2007; Fernandez-Duque et al. 2003; Koivisto et al. 2008; Lamy et al. 2009; Niedeggen et al. 2001; Pins and Ffytche 2003; Sergent et al. 2005). A similarly late effect is seen in MEG (van Aalderen-Smeets et al. 2006). The generators of the P3b are thought to involve a highly distributed set of simultaneously active areas, including hippocampus and temporal, parietal and frontal association cortices (Halgren et al. 1998; Mantini et al. 2009), that are proposed as components of the GNW.

7.3 *Intracranial Recordings*

The high spatial precision and signal-to-noise ratio afforded by intracranial recording in epileptic patients, combined with the human ability to perform objective and subjective perception tasks, provided a unique window into the neuronal correlates of conscious access (Fisch et al. 2009; Gaillard et al. 2009). Gaillard et al. (2009) contrasted the fate of masked (subliminal) versus unmasked (conscious) words while recording from a total of 176 local sites using intracortical depth electrodes in ten epileptic patients. Four objective measures of conscious perception were identified: (1) late (>300 ms) and distributed event-related potentials contacting sites in PFC; (2) large and late (>300 ms) increases in induced power (indexing local synchrony) in high-gamma frequencies (50–100 Hz), accompanied by a decrease in lower frequency power (centered on 10 Hz); (3) increases in long-distance cortico-cortical synchrony in the beta frequency band 13–30 Hz; and (4) increases in causal relations amongst distant cortical areas (Granger causality), bidirectionally but more strongly in the bottom–up direction.

Gaillard et al. (2009) noted that all four signatures coincided in the same time window (300–500 ms) and suggested that they might constitute different measures of the same state of distributed "ignition" of a large cortical network, including PFC. Compatible with the GNW concept of "global broadcasting", seen stimuli had a global impact on late evoked activity virtually anywhere in the cortex: 68.8% of electrode sites, although selected for clinical purposes, were modulated by the presence of conscious words (as opposed to 24.4% of sites for non-conscious words).

Intracranial recordings clearly refute an early speculation concerning the neural correlates of conscious access (Crick and Koch 1990): local gamma oscillations are *not* unique to conscious states, being clearly present on non-conscious trials, particularly during the first 200 ms of stimulus processing (Fisch et al. 2009; Gaillard et al. 2009; see also Melloni et al. 2007). Gamma-band power merely shows a relative increase on visible relative to invisible trials (Doesburg et al. 2009; Gaillard et al. 2009; Melloni et al. 2007; Rodriguez et al. 1999; Schurger et al. 2006; Wyart and Tallon-Baudry 2009). In recent visual intracranial recordings, for instance, gamma-band power increases were present in the 150–200 ms time window on non-consciously perceived trials but were significantly enhanced on consciously perceived trials, turning into an all-or-none pattern after 200 ms (see Fig. 6c in Fisch et al. 2009).

While time-frequency *power* increases occur in a very broad band (up to 100 Hz and above), in several intracranial MEG and EEG studies, *synchrony* across distant cortical sites tends to appear in the lower frequency range, either in the beta band 13–30 Hz or in the low-gamma band around 30–40 Hz, and to be a more specific indicator often exclusively seen during conscious access (Doesburg et al. 2009; Gaillard et al. 2009; Gross et al. 2004; Kranczioch et al. 2007; Rodriguez et al. 1999). With their slow period (25–80 ms), these frequencies may be more appropriate for forming brain-scale assemblies across long cortical distances and time delays (Fries 2005).

7.4 Neuronal Recordings

Animal single- and multi-unit recordings have examined changes in spike trains evoked by detected or undetected stimuli. These studies have largely supported and enhanced the conclusions drawn from human neuroimaging studies, and particularly the distinction between an early time window, dominated by stimulus-induced non-conscious processing, and a later window, dependent on long-distance interactions with higher association areas and reflecting conscious perception (Lamme and Roelfsema 2000).

A pioneering research program was conducted by Logothetis and collaborators using monkeys trained to report their perception during binocular rivalry (Leopold and Logothetis 1996; Sheinberg and Logothetis 1997). By recording from V1, V2, V4, MT, MST, IT and STS neurons and presenting two rivalling images, only one of which led to high neural firing, they identified a fraction of cells whose firing rate increased when their preferred stimuli was perceived, thus participating in a conscious neuronal assembly. The proportion of such cells increased from 20% in V1/V2 to 40% in V4, MT or MST, to as high as 90% in IT and STS. This finding supports the hypothesis that subjective perception is associated with distributed cell assemblies whose neurons are denser in higher associative cortices than in primary and secondary visual cortices.

Other paradigms afforded a precise measurement of the timing of information progression and conscious access in the visual system. In area V1, multi-unit recordings during both threshold judgments (Super et al. 2001) and masking paradigms (Lamme et al. 2002) identified two successive response periods. The first period (approximately 30–80 ms) was phasic, time-locked to stimulus onset, and reflected objective properties such as stimulus orientation, whether or not they were detectable by the animal. The second period, starting at 80–160 ms and extending for several hundreds of milliseconds, was associated with a late, slow and long-lasting amplification of firing rate that correlated tightly with stimulus detectability. Only this late amplification vanished under anesthesia (Lamme et al. 1998). Thus, although different forms of masking can affect both initial neural responses and late sustained discharges, in both awake and anesthesized monkeys (Macknik and Haglund 1999; Macknik and Livingstone 1998), the work of Lamme and colleagues suggests that it is the late phase that is most systematically correlated with conscious visibility.

In the infero-temporal cortex (IT), similarly, even when an image was presented extremely briefly and masked at a very short time interval (20–40 ms, thus probably subliminal), neurons still showed a brief pulse of firing, extending ~ 60–160 ms post-stimulus and carrying significant information about the masked image (Kovacs et al. 1995; Rolls et al. 1999). However, the mask image interrupted the firing train and prevented the emergence of a late discharge period (>200 ms post-stimulus), during which stimulus information was amplified.

Perhaps the clearest evidence for a two-stage process was obtained in the FEF in two macaque monkeys trained to detect faint and masked stimuli (Thompson and Schall 1999, 2000). A signal-detection analysis was applied, separating hits, misses, false alarms and correct rejections. A first firing peak (~80–100 ms post-stimulus), locked to stimulus onset, mostly tracked objective stimulus presence, although it was partially modulated according the monkey's eventual decision (i.e., higher on hits than on misses, and on false alarms than on correct rejections). A subsequent slow-firing modulation, starting around 100 ms and lasting several hundreds of milliseconds and thus bridging between visual and motor responses, was present only when monkeys detected a target, as indexed by their ocular responses.

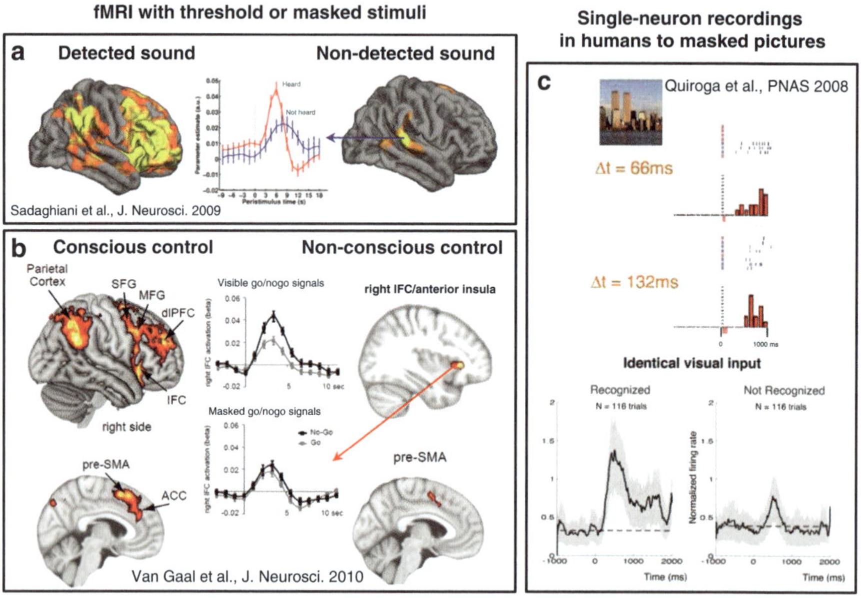

Fig. 4 Converging evidence from multiple methodologies for a global ignition during conscious access. (**a**) fMRI study of threshold-level noises, approximately half of which were consciously detected (Sadaghiani et al. 2009). Bilateral auditory areas showed a non-conscious activation that was amplified and spread to distributed inferior parietal, prefrontal and cingulate areas (for similar results with tactile stimuli, see Boly et al. 2007). (**b**) fMRI study of conscious versus non-conscious inhibitory control (van Gaal et al. 2010). Subjects were presented with masked visual signals, at the threshold for conscious perception, some of which occasionally required inhibiting a response (go/nogo task). Small activations to the non-conscious no-go signal were detected in the inferior frontal and pre-SMA cortices, but inhibitory control by a conscious no-go signal was associated with fMRI signal amplification (see the difference between no-go and go signals in middle graphs) and massive spread of the activation to additional and more anterior areas including prefrontal, anterior cingulate and inferior parietal cortices. (**c**) Single-cell correlate of all-or-none conscious access for masked pictures (Quiroga et al. 2008). Neurons were recorded from the human medial temporal lobe and hippocampus during presentation of masked pictures, with a variable target-mask delay. A single cell that fired specifically to pictures of the World Trade Center and did so only on trials when the patient recognized the picture (*dark blue* raster plots) and not on trials when recognition failed (*red* raster plots). In the plot of firing rate averaged across all neurons (bottom), transient firing could be seen on unrecognized trials, and conscious perception was characterized by a massive and durable amplification of activity (for complementary results in occipito-temporal areas, see also Fisch et al. 2009)

Thompson and Schall note that FEF neurons project back to extrastriate cortex and speculate that, “this top–down modulation (. . .) is a critical event to make explicit the interpretation of the ambiguous sensory input.”

Only a single study to date has explored single-neuron responses to seen or unseen stimuli in human cortex (Quiroga et al. 2008). Pictures followed at a variable delay by a mask were presented while recording from the medial temporal lobe in five patients with epilepsy. In this hierarchically high area, a very late response was seen, peaking around 300 ms and extending for 1 or 2 s. This late firing reflected tightly the person's subjective report to such an extent that individual trials reported as seen or unseen could be categorically distinguished by the neuron's firing train (see Fig. 4). Such a late categorical response is consistent with the hypothesis that conscious access is "all-or-none", leading either to a high degree of reverberation in higher association cortex or to a quickly decaying response (Sergent et al. 2005; Sergent and Dehaene 2004). Zhang and Luck (2008, 2009) have similarly observed that working memory operates in an all-or-none manner: only up to three "slots" are available, and manipulations that degrade memory, such as set size or masking of the targets, merely change the storage probability, not the number or precision of the memory.

Single-cell electrophysiology has also contributed to a better description of the mechanisms of formation of global brain-scale assemblies, confirming the key role of synchronization in conscious access. Within a single area such as V4, the degree to which single neurons synchronize with the ongoing fluctuations in local-field potential is a predictor of stimulus detection (Womelsdorf et al. 2006). Across distant areas such as FEF and V4 (Gregoriou et al. 2009) or PFC and LIP (Buschman and Miller 2007), synchrony is enhanced when the stimulus in the receptive field is attended and is thus presumably accessed consciously. Consistent with human MEG and intracranial studies (e.g., Gaillard et al. 2009; Gross et al. 2004), synchronization involves both gamma and beta bands, the latter being particularly enhanced during top–down attention (Buschman and Miller 2007). During the late phase of attention-driven activity, causal relations between distant areas are durably enhanced in both directions but more strongly so in the bottom–up direction from V4 to FEF (Gregoriou et al. 2009), again similar to human findings (Gaillard et al. 2009) and compatible with the idea that sensory information needs to be propagated anteriorly, particularly to PFC, before becoming consciously reportable.

8 Clinical Application to Coma, Vegetative State and Non-Communicative Disorders

The exploration of patients suffering from disorders of consciousness (DOC patients) such as comatose, vegetative state (VS) or other related states is motivated by two major goals. First, such translational research could eventually lead to major medical improvements in the way we (1) diagnose consciousness in non-communicating patients prognosis, (2) predict recovery of consciousness and cognitive abilities, (3) enable a bidirectional communication in these patients, and (4) facilitate recovery from impairments of arousal and consciousness. Second, these studies on such extreme situations offer a unique opportunity to test crucial theoretical predictions of our theories and to discover additional neural mechanisms involved in the physiology of consciousness. One such ancient but major discovery is the dissociation between arousal and consciousness as observed, for instance, in the vegetative state (see below). We will now briefly describe the various states observed in DOC patients and in general anesthesia and review some of the major findings related to this field.

8.1 Varieties of Disorders of Consciousness

Coma, from the Greek "deep sleep", is a drastic loss of responsivity, eye opening, visual pursuit and communicative acts, yet with a preservation of several reflexes. It is characterized by a failure of the arousal system caused by a structural or metabolic lesion of the brainstem reticular system or by widespread bilateral cerebral damage (Tshibanda et al. 2009, 2010). Coma is usually a transient condition that lasts no longer than a few days or weeks and evolves either to brain death, vegetative state, or regaining of consciousness. Contrary to coma patients, vegetative state (VS) patients exhibit preserved autonomous functioning, including spontaneous or stimulus-induced arousal and sleep–wake cycles (Laureys 2005). They may cry, smile, or move their eyes, head and limbs but in a meaningless, "automatic" manner and without any apparent voluntary or conscious response to verbal cues. Minimal consciousness (MCS), finally, is characterized by occasional but often unreliable voluntary responses to verbal commands or goal-driven eye movements (Giacino 2005). Unfortunately, the borders between these categories remain poorly defined, and even fully conscious but paralyzed patients with locked-in syndrome can remain undetected. It is therefore of interest to see if objective neural measures and GNW theory can help discriminate them.

8.2 Testing GNW Theory in Disorders of Consciousness

In coma and VS, as with general anesthesia, global metabolic activity typically decreases to ~50% of normal levels (Laureys 2005). This decrease is not homogeneous, however, but particularly pronounced in GNW areas including lateral and mesial prefrontal and inferior parietal cortices – a network that strongly reactivates, together with thalamo-cortical activity, upon recovery from VS (Laureys et al. 1999, 2000). Anatomically, prediction of recovery from coma relies on the comprehensive assessment of all structures involved in arousal and awareness functions, namely, the ascending reticular activating system located in the postero-superior part of the brainstem and structures encompassing thalamus, basal forebrain, and fronto-parietal association cortices (Tshibanda et al. 2009, 2010). Lesion or inhibition of part of this system suffices to cause immediate coma (e.g., Parvizi and Damasio 2003). Studies on traumatic coma patients with conventional MRI showed that lesions of the pons, midbrain, and basal ganglia were predictive of poor outcome, especially when they were bilateral (Tshibanda et al. 2010). In relation with the GNW model, it is noteworthy that prediction of non-recovery after 1 year could be calculated with up to 86% sensitivity and 97% specificity when taking into account both diffusion tensor and spectroscopic measures of cortical white matter integrity (Tshibanda et al. 2009).

As concerns the processing of external stimuli, GNW theory predicts that the initial but not the late stages of processing may be preserved in coma and VS patients. Indeed, several fMRI activation studies and $H_2{}^{15}O$ positron emission tomography (PET) studies in VS patients have shown that high-intensity electrical stimulation (experienced as painful in controls) still causes robust poststimulus activation restricted to brainstem, thalamus and primary somatosensory cortex (Laureys et al. 2002a; Tshibanda et al. 2010). Using fMRI, auditory, visual, tactile and pain-induced activations have been observed (Laureys et al. 2005; Owen et al. 2005). However, sensory areas appear to be functionally disconnected from "higher order" associative multimodal areas that fail to activate, such as, in the case of pain, secondary somatosensory, insular, posterior parietal and anterior cingulate cortices (Boly et al. 2008; Laureys et al. 2002a). The VS may thus be interpreted as resulting from a functional disconnection from the GNW circuits.

Two major predictions of GNW theory deal with the non-linearity of consciousness. First, we predict transitions between conscious and non-conscious states (intransitive consciousness) to be all-or-none in response to linear progressive increases (or decreases) of arousal, and second we predict that, in a conscious subject, access to an initially non-conscious representation should follow a sudden non-linear ignition process (conscious access). A recent study (Velly et al. 2007) recorded both cortical (scalp EEG) and subcortical (intracerebral local field potentials, LFP) activities during propofol anesthesia in Parkinsonian patients implanted with deep intracerebral electrodes. While subcortical structures showed a progressive linear decrease of the complexity (dimensional activation) of neural activity in response to progressive increase of propofol, cortical activity showed a brutal decrease of complexity. In the same vein, Ferrarelli et al. (2010) used an elegant EEG/TMS approach to demonstrate that, while a single TMS cortical pulse caused both a local early response and a late distributed pattern of cortical activity in distant areas including in particular frontal regions, the very same stimulations only caused the early local response when subjects had lost consciousness secondary to midazolam anesthesia. In VS patients, Laureys et al. (2002b) were the first to show that, while nociceptive stimuli could cause local patterns of cortical activation in somatoensory areas, only conscious subjects showed long-distance patterns of cortical activity with a strong functional connectivity between sensory areas and fronto-parietal cortices. Recently, it was also reported that loss of consciousness observed during epileptic seizures originating from the temporal lobe is a brutal event associated with a sharp increase of long-distance cortico-cortical and cortico-thalamic synchronization (Arthuis et al. 2009).

8.3 Clinical Application of the Signatures of Conscious Access

Can the observed signatures of conscious access in normal subjects be used to improve our ability to probe consciousness in DOC patients in whom behavioral manifestations are very limited? The existence of various forms of non-conscious patterns of cortical activity raises a challenging issue: how can we separate conscious and non-conscious patterns of brain activity? Progress in the cognitive neuroscience of conscious processing offers the hope of using markers of brain activity constrained by the specific properties of conscious processing to address this issue.

Using fMRI, a patient initially classified as vegetative by clinical signs showed essentially normal activations of distributed long-distance cortical networks during speech processing and mental imagery tasks (Owen et al. 2006). Recently, four more VS patients (out of 23) showed distinct cortical activation when asked to imagine playing tennis or navigating in a known place. Furthermore, one patient could voluntarily control the activation of these networks to provide yes/no answers to simple personal questions, clearly indicating some degree of preserved conscious processing (Monti et al. 2010). These important results confirm that some conscious patients are misclassified as "vegetative" and offer an empirical means of identifying them. However, they do so with a task whose relation to conscious processing is not founded in any precise theory.

In an effort to isolate such a theoretically validated scalp signature of conscious sensory processing, Bekinschtein et al. (2009) recorded ERPs to local versus global violations of an auditory regularity. As previously demonstrated (e.g., Fischer et al. 2004), when hearing a deviant tone after a sequence of repeated standard tones (sequence XXXXY), a local mismatch response was elicited non-consciously even in coma and VS patients. When this sequence XXXXY was repeatedly presented, however, such that the final tone change could be expected, the presentation of a deviant monotonic sequence (XXXXX) engendered a P3b wave in normal subjects that was absent in coma patients and in most VS patients but could still be observed in minimally conscious and locked-in patients. This paradigm, founded upon previous identification of the P3b component as a signature of conscious processing, still requires extended validation as a means of identifying residual conscious processing in patients. Recently, we reported the results of this test in a population of 31 VS patients. While most VS patients were negative on this test, we could identify 2 patients showing this neural signature of consciousness. Interestingly, these 2 patients showed unequivocal clinical signs of consciousness within the 3-4 days following ERP recording (Faugeras F et al. 2011).

Can the emerging understanding of conscious processing eventually propose novel therapeutic and pharmacological tools for patients suffering from impaired consciousness? Deep brain stimulation (DBS) in the thalamus, upper brain stem, and associated targets is advanced as a method to restore consciousness after loss of consciousness (LOC) due to severe brain injury (Cohadon and Richer 1993; Hassler et al. 1969a, b; Schiff et al. 2002, 2008; Sturm et al. 1979). Animal experimentation has demonstrated behavioral improvements when central thalamic stimulation is applied at the level of the anterior component of the intralaminar system, where neurons collect afferents from the brainstem arousal systems and connect to medial frontal cortices (Schiff et al. 2002). In a recent report (Schiff et al. 2008), electrodes were implanted bilaterally into the central thalamus of a single human subject who had remained in a minimally conscious state for 6 years following a deep traumatic brain injury. Over a 6-month period, communication, cognition, movement control and oral feeding improved during periods in which DBS was on as compared with periods in which it was off. This study is in great need of extension with more patients and with neuroimaging measures of recovery mechanisms. The GNW model predicts that improvements should relate to the reactivation of long-distance prefrontal and parietal networks, not just local thalamic effects. In VS, indeed, spontaneous recovery is accompanied by a functional restoration of a very broad frontoparietal network (Laureys et al. 1999) and some of its cortico-thalamo-cortical connections (Laureys et al. 2000). Furthermore, DTI (diffusion tensor imaging) in a patient who had been in a minimally conscious state but recovered verbal communication nearly two decades after a traumatic brain injury revealed an unanticipated axonal regrowth (Voss et al. 2006).

9 Conclusions

The project of relating subjective reports of conscious perception to objective behavioral and neuroscientific findings is now under way in many laboratories throughout the world. The GNW hypothesis provides one possible coherent framework within which these disparate observations can be integrated. It is exciting to see that this research program not only leads to novel experimental predictions in normal subjects but also, most importantly, to new clinical tools. Proving the efficacy of this approach in a broader group of patients will be a crucial endeavor for the coming years.

Acknowledgment This chapter is based in part on an earlier draft of a review article by S. Dehaene and J.P. Changeux, entitled "Experimental and Theoretical Approaches to Conscious Processing", to appear in Neuron in May 2011. We gratefully acknowledge extensive discussions with Sid Kouider, Jérôme Sackur, Bechir Jarraya, and Pierre-Marie Lledo as well as commens on previous drafts by Stuart Edelstein, Raphaël Gaillard, Biyu He and Henri Korn. This work was supported by Collège de France, INSERM, CNRS, European Research Council and Human Frontiers Science Program (S.D.), and Skaggs Research Foundation at UCSD School of Pharmacy (J.P.C.).

References

Alkire MT, Hudetz AG, Tononi G (2008) Consciousness and anesthesia. Science 322:876–880

Arthuis M, Valton L, Régis J, Chauvel P, Wendling F, Naccache L, Bernard C, Bartolomei F (2009) Impaired consciousness during temporal lobe seizures is related to increased long-distance cortical-subcortical synchronization. Brain 132:2091–2101

Baars BJ (1989) A cognitive theory of consciousness. Cambridge University Press, Cambridge, MA

Babiloni C, Vecchio F, Miriello M, Romani GL, Rossini PM (2006) Visuo-spatial consciousness and parieto-occipital areas: a high-resolution EEG study. Cereb Cortex 16:37–46

Balkin TJ, Braun AR, Wesensten NJ, Jeffries K, Varga M, Baldwin P, Belenky G, Herscovitch P (2002) The process of awakening: a PET study of regional brain activity patterns mediating the re-establishment of alertness and consciousness. Brain 125:2308–2319

Barbas H, Pandya DN (1989) Architecture and intrinsic connections of the prefrontal cortex in the rhesus monkey. J Comp Neurol 286:353–375
Bassett DS, Bullmore ET (2009) Human brain networks in health and disease. Curr Opin Neurol 22:340–347
Bassett DS, Bullmore E, Verchinski BA, Mattay VS, Weinberger DR, Meyer-Lindenberg A (2008) Hierarchical organization of human cortical networks in health and schizophrenia. J Neurosci 28:9239–9248
Bassett DS, Bullmore ET, Meyer-Lindenberg A, Apud JA, Weinberger DR, Coppola R (2009) Cognitive fitness of cost-efficient brain functional networks. Proc Natl Acad Sci USA 106: 11747–11752
Bekinschtein TA, Dehaene S, Rohaut B, Tadel F, Cohen L, Naccache L (2009) Neural signature of the conscious processing of auditory regularities. Proc Natl Acad Sci USA 106:1672–1677
Boly M, Balteau E, Schnakers C, Degueldre C, Moonen G, Luxen A, Phillips C, Peigneux P, Maquet P, Laureys S (2007) Baseline brain activity fluctuations predict somatosensory perception in humans. Proc Natl Acad Sci USA 104:12187–12192
Boly M, Faymonville ME, Schnakers C, Peigneux P, Lambermont B, Phillips C, Lancellotti P, Luxen A, Lamy M, Moonen G, Maquet P, Laureys S (2008) Perception of pain in the minimally conscious state with PET activation: an observational study. Lancet Neurol 7:1013–1020
Bourgeois JP, Goldman-Rakic PS, Rakic P (1994) Synaptogenesis in the prefrontal cortex of rhesus monkeys. Cereb Cortex 4:78–96
Bridgeman B (1975) Correlates of metacontrast in single cells of the cat visual system. Vision Res 15:91–99
Bridgeman B (1988) Visual evoked potentials: concomitants of metacontrast in late components. Percept Psychophys 43:401–403
Bullmore E, Sporns O (2009) Complex brain networks: graph theoretical analysis of structural and functional systems. Nat Rev Neurosci 10:186–198
Buschman TJ, Miller EK (2007) Top-down versus bottom-up control of attention in the prefrontal and posterior parietal cortices. Science 315:1860–1862
Cajal SR, DeFelipe J, Jones EG (1892/1988) Cajal on the cerebral cortex: an annotated translation of the complete writings. History of Neuroscience, No 1. New York: Oxford University Press
Catani M, Thiebaut de Schotten M (2008) A diffusion tensor imaging tractography atlas for virtual in vivo dissections. Cortex 44:1105–1132
Cavada C, Company T, Tejedor J, Cruz-Rizzolo RJ, Reinoso-Suarez F (2000) The anatomical connections of the macaque monkey orbitofrontal cortex. A review. Cereb Cortex 10:220–242
Cavanna AE, Trimble MR (2006) The precuneus: a review of its functional anatomy and behavioural correlates. Brain 129:564–583
Changeux JP, Danchin A (1976) Selective stabilization of developing synapses as a mechanism for the specification of neuronal networks. Nature 264:705–712
Changeux JP, Dehaene S (1989) Neuronal models of cognitive functions. Cognition 33:63–109
Changeux JP, Courrège P, Danchin A (1973) A theory of the epigenesis of neural networks by selective stabilization of synapses. Proc Natl Acad Sci USA 70:2974–2978
Christoff K, Gordon AM, Smallwood J, Smith R, Schooler JW (2009) Experience sampling during fMRI reveals default network and executive system contributions to mind wandering. Proc Natl Acad Sci USA 106:8719–8724
Chun MM, Potter MC (1995) A two-stage model for multiple target detection in rapid serial visual presentation. J Exp Psychol Hum Percept Perform 21:109–127
Cohadon F, Richer E (1993) Deep cerebral stimulation in patients with post-traumatic vegetative state: 25 cases. Neurochirurgie 39:281–292
Crick F, Koch C (1990) Toward a neurobiological theory of consciousness. Sem Neurosci 2:263–275
Crick F, Koch C (2003) A framework for consciousness. Nat Neurosci 6:119–126
Damasio A (1999) The feeling of what happens. Harcourt Brace and Co, New York

de Lafuente V, Romo R (2006) Neural correlate of subjective sensory experience gradually builds up across cortical areas. Proc Natl Acad Sci USA 103:14266–14271

Deco G, Perez-Sanagustin M, de Lafuente V, Romo R (2007) Perceptual detection as a dynamical bistability phenomenon: a neurocomputational correlate of sensation. Proc Natl Acad Sci USA 104:20073–20077

DeFelipe J, Farinas I (1992) The pyramidal neuron of the cerebral cortex: morphological and chemical characteristics of the synaptic inputs. Prog Neurobiol 39:563–607

Dehaene S, Changeux JP (1989) A simple model of prefrontal cortex function in delayed-response tasks. J Cogn Neurosci 1:244–261

Dehaene S, Changeux JP (1991) The Wisconsin Card Sorting Test: theoretical analysis and modelling in a neuronal network. Cereb Cortex 1:62–79

Dehaene S, Changeux JP (1997) A hierarchical neuronal network for planning behavior. Proc Natl Acad Sci USA 94:13293–13298

Dehaene S, Changeux JP (2005) Ongoing spontaneous activity controls access to consciousness: a neuronal model for inattentional blindness. PLoS Biol 3(5):e141

Dehaene S, Naccache L (2001) Towards a cognitive neuroscience of consciousness: Basic evidence and a workspace framework. Cognition 79:1–37

Dehaene S, Kerszberg M, Changeux JP (1998) A neuronal model of a global workspace in effortful cognitive tasks. Proc Natl Acad Sci USA 95:14529–14534

Dehaene S, Naccache L, Cohen L, Le Bihan D, Mangin JF, Poline JB, Rivière D (2001) Cerebral mechanisms of word masking and unconscious repetition priming. Nat Neurosci 4:752–758

Dehaene S, Sergent C, Changeux JP (2003) A neuronal network model linking subjective reports and objective physiological data during conscious perception. Proc Natl Acad Sci USA 100:8520–8525

Dehaene S, Changeux JP, Naccache L, Sackur J, Sergent C (2006) Conscious, preconscious, and subliminal processing: a testable taxonomy. Trends Cogn Sci 10:204–211

Dejerine J (1895) Anatomie des centres nerveux, vol 1. Rueff et Cie, Paris

Del Cul A, Dehaene S, Leboyer M (2006) Preserved subliminal processing and impaired conscious access in schizophrenia. Arch Gen Psychiatr 63:1313–1323

Del Cul A, Baillet S, Dehaene S (2007) Brain dynamics underlying the nonlinear threshold for access to consciousness. PLoS Biol 5(10):e260

Del Cul A, Dehaene S, Reyes P, Bravo E, Slachevsky A (2009) Causal role of prefrontal cortex in the threshold for access to consciousness. Brain 132:2531–2540

Doesburg SM, Green JJ, McDonald JJ, Ward LM (2009) Rhythms of consciousness: binocular rivalry reveals large-scale oscillatory network dynamics mediating visual perception. PLoS One 4(7):e6142

Edelman GM (1989) The remembered present. Basic Books, New York

Elston GN (2003) Cortex, cognition and the cell: new insights into the pyramidal neuron and prefrontal function. Cereb Cortex 13:1124–1138

Elston GN, Rosa MG (1997) The occipitoparietal pathway of the macaque monkey: comparison of pyramidal cell morphology in layer III of functionally related cortical visual areas. Cereb Cortex 7:432–452

Elston GN, Rosa MG (1998) Morphological variation of layer III pyramidal neurones in the occipitotemporal pathway of the macaque monkey visual cortex. Cereb Cortex 8:278–294

Fahrenfort JJ, Scholte HS, Lamme VA (2007) Masking disrupts reentrant processing in human visual cortex. J Cogn Neurosci 19:1488–1497

Faugeras F, Rohaut B, Weiss N, Bekinschtein TA, Galanaud D, Puybasset L, Bolgert F, Sergent C, Cohen L, Dehaene S, Naccache N (2011) Probing consciousness with event-related potentials in the vegetative state. Neurology, doi: 10.1212/WNL.0b013e3182217ee8

Fernandez-Duque D, Grossi G, Thornton IM, Neville HJ (2003) Representation of change: separate electrophysiological markers of attention, awareness, and implicit processing. J Cogn Neurosci 15:491–507

Ferrarelli F, Massimini M, Sarasso S, Casali A, Riedner BA, Angelini G, Tononi G, Pearce RA (2010) Breakdown in cortical effective connectivity during midazolam-induced loss of consciousness. Proc Natl Acad Sci USA 107:2681–2686
Fisch L, Privman E, Ramot M, Harel M, Nir Y, Kipervasser S, Andelman F, Neufeld MY, Kramer U, Fried I, Malach R (2009) Neural "ignition": enhanced activation linked to perceptual awareness in human ventral stream visual cortex. Neuron 64:562–574
Fischer C, Luaute J, Adeleine P, Morlet D (2004) Predictive value of sensory and cognitive evoked potentials for awakening from coma. Neurology 63:669–673
Fries P (2005) A mechanism for cognitive dynamics: neuronal communication through neuronal coherence. Trends Cogn Sci 9:474–480
Fuster JM (2008) The prefrontal cortex, 4th edn. Academic, London
Gaillard R, Dehaene S, Adam C, Clémenceau S, Hasboun D, Baulac M, Cohen L, Naccache L (2009) Converging intracranial markers of conscious access. PLoS Biol 7(3):e61
Giacino JT (2005) The minimally conscious state: defining the borders of consciousness. Prog Brain Res 150:381–395
Goldman-Rakic PS (1988) Topography of cognition: Parallel distributed networks in primate association cortex. Annu Rev Neurosci 11:137–156
Goldman-Rakic PS (1996) The prefrontal landscape: implications of functional architecture for understanding human mentation and the central executive. Philos Trans R Soc Lond B Biol Sci 351:1445–1453
Goldman-Rakic PS (1999) The "psychic" neuron of the cerebral cortex. Ann N Y Acad Sci 868:13–26
Gong G, He Y, Concha L, Lebel C, Gross DW, Evans AC, Beaulieu C (2009) Mapping anatomical connectivity patterns of human cerebral cortex using in vivo diffusion tensor imaging tractography. Cereb Cortex 19:524–536
Gregoriou GG, Gotts SJ, Zhou H, Desimone R (2009) High-frequency, long-range coupling between prefrontal and visual cortex during attention. Science 324:1207–1210
Grill-Spector K, Kushnir T, Hendler T, Malach R (2000) The dynamics of object-selective activation correlate with recognition performance in humans. Nat Neurosci 3:837–843
Gross J, Schmitz F, Schnitzler I, Kessler K, Shapiro K, Hommel B, Schnitzler A (2004) Modulation of long-range neural synchrony reflects temporal limitations of visual attention in humans. Proc Natl Acad Sci USA 101:13050–13055
Gusnard DA, Raichle ME (2001) Searching for a baseline: functional imaging and the resting human brain. Nat Rev Neurosci 2:685–694
Hagmann P, Cammoun L, Gigandet X, Meuli R, Honey CJ, Wedeen VJ, Sporns O (2008) Mapping the structural core of human cerebral cortex. PLoS Biol 6(7):e159
Halgren E, Marinkovic K, Chauvel P (1998) Generators of the late cognitive potentials in auditory and visual oddball tasks. Electroencephalogr Clin Neurophysiol 106:156–164
Hassler R, Ore GD, Bricolo A, Dieckmann G, Dolce G (1969a) EEG and clinical arousal induced by bilateral long-term stimulation of pallidal systems in traumatic vigil coma. Electroencephalogr Clin Neurophysiol 27:689–690
Hassler R, Ore GD, Dieckmann G, Bricolo A, Dolce G (1969b) Behavioural and EEG arousal induced by stimulation of unspecific projection systems in a patient with post-traumatic apallic syndrome. Electroencephalogr Clin Neurophysiol 27:306–310
Haynes JD, Driver J, Rees G (2005) Visibility reflects dynamic changes of effective connectivity between V1 and fusiform cortex. Neuron 46:811–821
He BJ, Raichle ME (2009) The fMRI signal, slow cortical potential and consciousness. Trends Cogn Sci 13:302–309
He Y, Dagher A, Chen Z, Charil A, Zijdenbos A, Worsley K, Evans A (2009) Impaired small-world efficiency in structural cortical networks in multiple sclerosis associated with white matter lesion load. Brain 132:3366–3379
Iturria-Medina Y, Sotero RC, Canales-Rodriguez EJ, Aleman-Gomez Y, Melie-Garcia L (2008) Studying the human brain anatomical network via diffusion-weighted MRI and Graph Theory. Neuroimage 40:1064–1076

James W (1890) The principles of psychology. Holt, New York
Kim CY, Blake R (2005) Psychophysical magic: rendering the visible 'invisible'. Trends Cogn Sci 9:381–388
Koivisto M, Revonsuo A, Lehtonen M (2006) Independence of visual awareness from the scope of attention: an electrophysiological study. Cereb Cortex 16:415–424
Koivisto M, Lahteenmaki M, Sorensen TA, Vangkilde S, Overgaard M, Revonsuo A (2008) The earliest electrophysiological correlate of visual awareness? Brain Cogn 66:91–103
Koivisto M, Kainulainen P, Revonsuo A (2009) The relationship between awareness and attention: evidence from ERP responses. Neuropsychologia 47:2891–2899
Kouider S, Dehaene S, Jobert A, Le Bihan D (2007) Cerebral bases of subliminal and supraliminal priming during reading. Cereb Cortex 17:2019–2029
Kovacs G, Vogels R, Orban GA (1995) Cortical correlate of pattern backward masking. Proc Natl Acad Sci USA 92:5587–5591
Kranczioch C, Debener S, Maye A, Engel AK (2007) Temporal dynamics of access to consciousness in the attentional blink. Neuroimage 37:947–955
Kringelbach ML, Rolls ET (2004) The functional neuroanatomy of the human orbitofrontal cortex: evidence from neuroimaging and neuropsychology. Prog Neurobiol 72:341–372
Kritzer MF, Goldman-Rakic PS (1995) Intrinsic circuit organization of the major layers and sublayers of the dorsolateral prefrontal cortex in the rhesus monkey. J Comp Neurol 359:131–143
Lamme VA, Roelfsema PR (2000) The distinct modes of vision offered by feedforward and recurrent processing. Trends Neurosci 23:571–579
Lamme VA, Zipser K, Spekreijse H (1998) Figure-ground activity in primary visual cortex is suppressed by anesthesia. Proc Natl Acad Sci USA 95:3263–3268
Lamme VA, Zipser K, Spekreijse H (2002) Masking interrupts figure-ground signals in V1. J Cogn Neurosci 14:1044–1053
Lamy D, Salti M, Bar-Haim Y (2009) Neural correlates of subjective awareness and unconscious processing: an ERP study. J Cogn Neurosci 21:1435–1446
Lau HC, Passingham RE (2006) Relative blindsight in normal observers and the neural correlate of visual consciousness. Proc Natl Acad Sci USA 103:18763–18768
Laureys S (2005) The neural correlate of (un)awareness: lessons from the vegetative state. Trends Cogn Sci 9:556–559
Laureys S, Lemaire C, Maquet P, Phillips C, Franck G (1999) Cerebral metabolism during vegetative state and after recovery to consciousness. J Neurol Neurosurg Psychiatr 67:121
Laureys S, Faymonville ME, Luxen A, Lamy M, Franck G, Maquet P (2000) Restoration of thalamocortical connectivity after recovery from persistent vegetative state. Lancet 355:1790–1791
Laureys S, Antoine S, Boly M, Elincx S, Faymonville ME, Berré J, Sadzot B, Ferring M, De Tiège X, van Bogaert P, Hansen I, Damas P, Mavroudakis N, Lambermont B, Del Fiore G, Aerts J, Degueldre C, Phillips C, Franck G, Vincent JL, Lamy M, Luxen A, Moonen G, Goldman S, Maquet P (2002a) Brain function in the vegetative state. Acta Neurol Belg 102:177–185
Laureys S, Faymonville ME, Peigneux P, Damas P, Lambermont B, Del Fiore G, Degueldre C, Aerts J, Luxen A, Franck G, Lamy M, Moonen G, Maquet P (2002b) Cortical processing of noxious somatosensory stimuli in the persistent vegetative state. Neuroimage 17:732–741
Laureys S, Perrin F, Schnakers C, Boly M, Majerus S (2005) Residual cognitive function in comatose, vegetative and minimally conscious states. Curr Opin Neurol 18:726–733
Leopold DA, Logothetis NK (1996) Activity changes in early visual cortex reflect monkeys' percepts during binocular rivalry. Nature 379:549–553
Mack A, Rock I (1998) Inattentional blindness. MIT, Cambridge, MA
Macknik SL, Haglund MM (1999) Optical images of visible and invisible percepts in the primary visual cortex of primates. Proc Natl Acad Sci USA 96:15208–15210

Macknik SL, Livingstone MS (1998) Neuronal correlates of visibility and invisibility in the primate visual system. Nat Neurosci 1:144–149
Mantini D, Corbetta M, Perrucci MG, Romani GL, Del Gratta C (2009) Large-scale brain networks account for sustained and transient activity during target detection. Neuroimage 44:265–274
Mason MF, Norton MI, Van Horn JD, Wegner DM, Grafton ST, Macrae CN (2007) Wandering minds: the default network and stimulus-independent thought. Science 315:393–395
Melchitzky DS, Sesack SR, Pucak ML, Lewis DA (1998) Synaptic targets of pyramidal neurons providing intrinsic horizontal connections in monkey prefrontal cortex. J Comp Neurol 390: 211–224
Melchitzky DS, Gonzalez-Burgos G, Barrionuevo G, Lewis DA (2001) Synaptic targets of the intrinsic axon collaterals of supragranular pyramidal neurons in monkey prefrontal cortex. J Comp Neurol 430:209–221
Melloni L, Molina C, Pena M, Torres D, Singer W, Rodriguez E (2007) Synchronization of neural activity across cortical areas correlates with conscious perception. J Neurosci 27: 2858–2865
Meyer K, Damasio A (2009) Convergence and divergence in a neural architecture for recognition and memory. Trends Neurosci 32:376–382
Monti MM, Vanhaudenhuyse A, Coleman MR, Boly M, Pickard JD, Tshibanda L, Owen AM, Laureys S (2010) Willful modulation of brain activity in disorders of consciousness. N Engl J Med 362:579–589
Muller-Gass A, Macdonald M, Schroger E, Sculthorpe L, Campbell K (2007) Evidence for the auditory P3a reflecting an automatic process: elicitation during highly-focused continuous visual attention. Brain Res 1170:71–78
Niedeggen M, Wichmann P, Stoerig P (2001) Change blindness and time to consciousness. Eur J Neurosci 14:1719–1726
Niswender CM, Conn PJ (2010) Metabotropic glutamate receptors: physiology, pharmacology, and disease. Annu Rev Pharmacol Toxicol 50:295–322
Owen AM, Coleman MR, Menon DK, Berry EL, Johnsrude IS, Rodd JM, Davis MH, Pickard JD (2005) Using a hierarchical approach to investigate residual auditory cognition in persistent vegetative state. Prog Brain Res 150:457–471
Owen AM, Coleman MR, Boly M, Davis MH, Laureys S, Pickard JD (2006) Detecting awareness in the vegetative state. Science 313:1402
Palva S, Linkenkaer-Hansen K, Naatanen R, Palva JM (2005) Early neural correlates of conscious somatosensory perception. J Neurosci 25:5248–5258
Pandya DN, Yeterian EH (1990) Prefrontal cortex in relation to other cortical areas in rhesus monkey: architecture and connections. Prog Brain Res 85:63–94
Parvizi J, Damasio AR (2003) Neuroanatomical correlates of brainstem coma. Brain 126: 1524–1536
Parvizi J, Van Hoesen GW, Buckwalter J, Damasio A (2006) Neural connections of the posteromedial cortex in the macaque. Proc Natl Acad Sci USA 103:1563–1568
Persaud N, McLeod P, Cowey A (2007) Post-decision wagering objectively measures awareness. Nat Neurosci 10:257–261
Petrides M, Pandya DN (2009) Distinct parietal and temporal pathways to the homologues of Broca's area in the monkey. PLoS Biol 7(8):e1000170
Pins D, Ffytche D (2003) The neural correlates of conscious vision. Cereb Cortex 13:461–474
Preuss TM, Goldman-Rakic PS (1991) Ipsilateral cortical connections of granular frontal cortex in the strepsirhine primate Galago, with comparative comments on anthropoid primates. J Comp Neurol 310:507–549
Pucak ML, Levitt JB, Lund JS, Lewis DA (1996) Patterns of intrinsic and associational circuitry in monkey prefrontal cortex. J Comp Neurol 376:614–630
Quiroga RQ, Mukamel R, Isham EA, Malach R, Fried I (2008) Human single-neuron responses at the threshold of conscious recognition. Proc Natl Acad Sci USA 105:3599–3604

Railo H, Koivisto M (2009) The electrophysiological correlates of stimulus visibility and metacontrast masking. Conscious Cogn 18:794–803
Rockstroh B, Müller M, Cohen R, Elbert T (1992) Probing the functional brain state during P300 evocation. J Psychophysiol 6:175–184
Rodriguez E, George N, Lachaux JP, Martinerie J, Renault B, Varela FJ (1999) Perception's shadow: long-distance synchronization of human brain activity. Nature 397:430–433
Rolls ET, Tovee MJ, Panzeri S (1999) The neurophysiology of backward visual masking: information analysis. J Cogn Neurosci 11:300–311
Rougier NP, Noelle DC, Braver TS, Cohen JD, O'Reilly RC (2005) Prefrontal cortex and flexible cognitive control: rules without symbols. Proc Natl Acad Sci USA 102:7338–7343
Sadaghiani S, Hesselmann G, Kleinschmidt A (2009) Distributed and antagonistic contributions of ongoing activity fluctuations to auditory stimulus detection. J Neurosci 29:13410–13417
Salisbury D, Squires NK, Ibel S, Maloney T (1992) Auditory event-related potentials during stage 2 NREM sleep in humans. J Sleep Res 1:251–257
Schiff ND, Plum F, Rezai AR (2002) Developing prosthetics to treat cognitive disabilities resulting from acquired brain injuries. Neurol Res 24:116–124
Schiff ND, Giacino JT, Kalmar K, Victor JD, Baker K, Gerber M, Fritz B, Eisenberg B, O'Connor J, Kobylarz EJ, Farris S, Machado A, McCagg C, Plum F, Fins JJ, Rezai AR (2008) Behavioural improvements with thalamic stimulation after severe traumatic brain injury. Nature 452:120
Schiller PH, Chorover SL (1966) Metacontrast: its relation to evoked potentials. Science 153: 1398–1400
Schurger A, Sher S (2008) Awareness, loss aversion, and post-decision wagering. Trends Cogn Sci 12:209–210, author reply 210
Schurger A, Cowey A, Tallon-Baudry C (2006) Induced gamma-band oscillations correlate with awareness in hemianopic patient GY. Neuropsychologia 44:1796–1803
Self MW, Super H, Roelfsema PR (2008) Feedforward and feedback visual processing utilise different glutamate receptors. Paper presented at the Society for Neuroscience: Abstracts
Sergent C, Dehaene S (2004) Is consciousness a gradual phenomenon? Evidence for an all-or-none bifurcation during the attentional blink. Psychol Sci 15:720–728
Sergent C, Baillet S, Dehaene S (2005) Timing of the brain events underlying access to consciousness during the attentional blink. Nat Neurosci 8:1391–1400
Shallice T (1988) From neuropsychology to mental structure. Cambridge University Press, Cambridge, UK
Sheinberg DL, Logothetis NK (1997) The role of temporal cortical areas in perceptual organization. Proc Natl Acad Sci USA 94:3408–3413
Sigman M, Dehaene S (2005) Parsing a cognitive task: a characterization of the mind's bottleneck. PLoS Biol 3(2):e37
Sigman M, Dehaene S (2008) Brain mechanisms of serial and parallel processing during dual-task performance. J Neurosci 28:7585–7598
Smallwood J, Beach E, Schooler JW, Handy TC (2008) Going AWOL in the brain: mind wandering reduces cortical analysis of external events. J Cogn Neurosci 20:458–469
Sturm V, Kuhner A, Schmitt HP, Assmus H, Stock G (1979) Chronic electrical stimulation of the thalamic unspecific activating system in a patient with coma due to midbrain and upper brain stem infarction. Acta Neurochir (Wien) 47:235–244
Super H, Spekreijse H, Lamme VA (2001) Two distinct modes of sensory processing observed in monkey primary visual cortex (V1). Nat Neurosci 4:304–310
Super H, van der Togt C, Spekreijse H, Lamme V (2003) Internal state of monkey primary visual cortex (V1) predicts figure-ground perception. J Neurosci 23:3407–3414
Thompson KG, Schall JD (1999) The detection of visual signals by macaque frontal eye field during masking. Nat Neurosci 2:283–288
Thompson KG, Schall JD (2000) Antecedents and correlates of visual detection and awareness in macaque prefrontal cortex. Vision Res 40:1523–1538

Tse PU, Martinez-Conde S, Schlegel AA, Macknik SL (2005) Visibility, visual awareness, and visual masking of simple unattended targets are confined to areas in the occipital cortex beyond human V1/V2. Proc Natl Acad Sci USA 102:17178–17183

Tshibanda L, Vanhaudenhuyse A, Galanaud D, Boly M, Laureys S, Puybasset L (2009) Magnetic resonance spectroscopy and diffusion tensor imaging in coma survivors: promises and pitfalls. Prog Brain Res 177:215–229

Tshibanda L, Vanhaudenhuyse A, Boly M, Soddu A, Bruno MA, Moonen G, Laureys S, Noirhomme Q (2010) Neuroimaging after coma. Neuroradiology 52:15–24

van Aalderen-Smeets SI, Oostenveld R, Schwarzbach J (2006) Investigating neurophysiological correlates of metacontrast masking with magnetoencephalography. Adv Cog Psychol 2:21–35

van Gaal S, Ridderinkhof KR, Scholte HS, Lamme VA (2010) Unconscious activation of the prefrontal no-go network. J Neurosci 30:4143–4150

Velly LJ, Rey MF, Bruder NJ, Gouvitsos FA, Witjas T, Regis JM, Peragut JC, Gouin FM (2007) Differential dynamic of action on cortical and subcortical structures of anesthetic agents during induction of anesthesia. Anesthesiology 107:202–212

Vogel EK, Machizawa MG (2004) Neural activity predicts individual differences in visual working memory capacity. Nature 428:748–751

Vogel EK, Luck SJ, Shapiro KL (1998) Electrophysiological evidence for a postperceptual locus of suppression during the attentional blink. J Exp Psychol Hum Percept Perform 24:1656–1674

Vogt BA, Laureys S (2005) Posterior cingulate, precuneal and retrosplenial cortices: cytology and components of the neural network correlates of consciousness. Prog Brain Res 150:205–217

Von Economo C (1929) The cytoarchitectonics of the human cerebral cortex. Oxford University Press, London

Voss HU, Uluç AM, Dyke JP, Watts R, Kobylarz EJ, McCandliss BD, Heier LA, Beattie BJ, Hamacher KA, Vallabhajosula S, Goldsmith SJ, Ballon D, Giacino JT, Schiff ND (2006) Possible axonal regrowth in late recovery from the minimally conscious state. J Clin Invest 116:2005–2011

Watts DJ, Strogatz SH (1998) Collective dynamics of 'small-world' networks. Nature 393: 440–442

Williams MA, Visser TA, Cunnington R, Mattingley JB (2008) Attenuation of neural responses in primary visual cortex during the attentional blink. J Neurosci 28:9890–9894

Womelsdorf T, Fries P, Mitra PP, Desimone R (2006) Gamma-band synchronization in visual cortex predicts speed of change detection. Nature 439:733–736

Wyart V, Tallon-Baudry C (2008) Neural dissociation between visual awareness and spatial attention. J Neurosci 28:2667–2679

Wyart V, Tallon-Baudry C (2009) How ongoing fluctuations in human visual cortex predict perceptual awareness: baseline shift versus decision bias. J Neurosci 29:8715–8725

Zhang W, Luck SJ (2008) Discrete fixed-resolution representations in visual working memory. Nature 453:233–235

Zhang W, Luck SJ (2009) Sudden death and gradual decay in visual working memory. Psychol Sci 20:423–428

Zylberberg A, Dehaene S, Mindlin GB, Sigman M (2009) Neurophysiological bases of exponential sensory decay and top-down memory retrieval: a model. Front Comput Neurosci 3:4

Zylberberg A, Fernandez Slezak D, Roelfsema PR, Dehaene S, Sigman M (2010) The brain's router: a cortical network model of serial processing in the primate brain. PLoS Comp Biol 6: e1000765

Disorders of Consciousness: What Do We Know?

Camille Chatelle, Steven Laureys, and Caroline Schnakers

Abstract Improving the diagnosis of altered states of consciousness is of ethical and medical interest. The alarming rate of misdiagnosis shows how difficult it is to disentangle disorders of consciousness. This review shows that the integration of objective methods such as neuroimaging techniques should improve our diagnoses and guide our clinical therapeutic options in these challenging patients. Additionally, just as resting metabolism and passive paradigms studies increased our understanding of residual cerebral processing in vegetative and minimally conscious patients, an active paradigm seems to be another powerful objective diagnostic tool to distinguish between these patients.

Differentiating reflex from voluntary activity is one of the most challenging tasks facing clinicians involved in the care of patients recovering from coma. Behavioral assessment is considered to be the main way to detect signs of consciousness and, hence, to determine a diagnosis (Majerus et al. 2005). At the bedside, patients in a vegetative state present eyes-opening and can sometimes show a few movements or reactions to stimulation. All these behaviors are considered to be reflexes and not to reflectconscious brain activity. However, behavioral assessment is complicated by the presence of motor impairment, tracheotomy, fluctuating arousal level or ambiguous and rapidly habituating responses (Gill-Thwaites 2006). Because of these compromising factors, diagnosis can rapidly become difficult to make. A recent study has shown a misdiagnosis of 41% in patients who were clinically diagnosed as being in a vegetative state (Schnakers et al. 2009a). To improve on this high misdiagnosis rate, we need standardized but also sensitive measures to detect signs of consciousness in these patients. Specialized neuro-behavioral rating scales have been developed to provide a reliable and valid means of detecting signs of

C. Schnakers (✉)
Coma Science Group, Cyclotron Research Centre, University of Liege, Sart Tilman, B30, 4000 Liège, Belgium
e-mail: C.Schnakers@ulg.ac.be

S. Dehaene and Y. Christen (eds.), *Characterizing Consciousness: From Cognition to the Clinic?* Research and Perspectives in Neurosciences,
DOI 10.1007/978-3-642-18015-6_5,

consciousness. However, these techniques are still influenced by the patients' motor abilities. As misdiagnosis can lead to grave consequences, especially in terms of pain treatment and end-of-life decision-making (Andrews 2004), additional tools have to be used to assess remnant brain functioning and detect brain activity linked to consciousness.

Functional neuroimaging techniques such as positron emission tomography (PET) and functional magnetic resonance imaging (fMRI) can provide an objective index of brain activity at rest and during active cognitive processing. These techniques are well equipped to identify covert cognitive processes in patients who are otherwise incapable of intelligible or sustained behavioral expression, and they offer complementary information to bedside examination findings.

1 Defining Altered States of Consciousness

Consciousness consists of two major components: arousal and awareness (Laureys et al. 2002a). Clinically, arousal is manifested by spontaneous eye opening. Awareness, on the other hand, can be assessed by responses to external stimuli (e.g., command following, visual pursuit). If awareness requires arousal, preserved arousal levels do not necessarily imply awareness. When a patient sustains severe brain damage, he can go through different states of altered consciousness before fully recovering consciousness (Fig. 1). These states can be described in terms of degree of arousal and awareness.

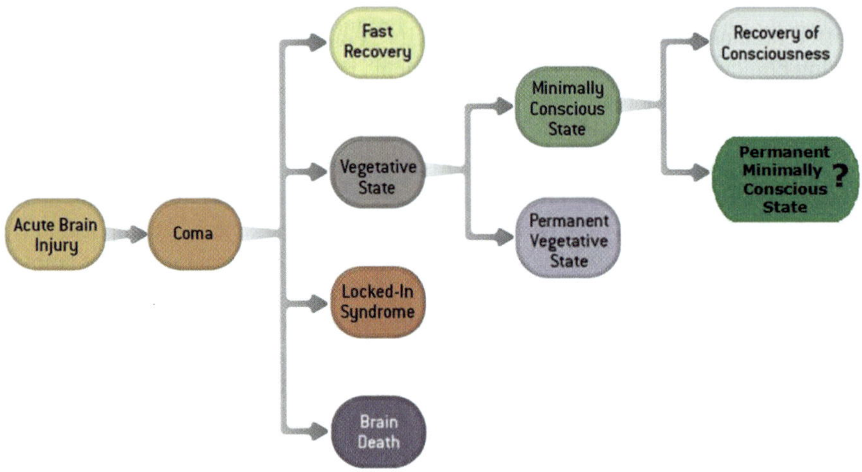

Fig. 1 From coma to consciousness. After a brain injury leading to coma, a patient's progress may follow one of several paths. If the patient does not die (brain death) or quickly recover, he or she will usually go through a vegetative state. In rare cases, he may develop a locked-in syndrome, a complete paralysis of the body's voluntary muscles. The patient may then evolve to the minimally conscious state, and often further recovery of consciousness, or remain in the vegetative state permanently [adapted from Laureys (2007)]

1.1 Brain Death

Brain death suggests that the organism cannot function as a whole. Critical functions such as respiration and circulation, neuroendocrine and homeostatic regulation, and consciousness are permanently absent. The patient is apneic and unreactive to environmental stimulation (Guidelines for the determination of death 1981). The term "brain death" requires the bedside demonstration of irreversible cessation of all clinical functions of the brain, but also, the brainstem. Brain death is classically caused by a brain lesion (for example, massive traumatic injury, intracranial haemorrhage or anoxia) that results in an intracranial pressure higher than the mean arterial blood pressure. After excluding the impact of pharmacological (or toxic) treatments or hypothermia, the diagnosis can be done within 6–24 h.

1.2 Coma

If they survive, patients can stay in a coma for several weeks, being neither aroused nor aware; indeed, their eyes are constantly closed and they do not manifest voluntary behavioral responses. Generally, patients emerge from their comatose state within 2–4 weeks (Posner et al. 2007). The prognosis is influenced by different factors such as etiology, the patient's general medical condition and age. Outcome is known to be bad if, after 3 days of observation, there are no pupillary or corneal reflexes, stereotyped or absent motor response to noxious stimulation, or iso-electrical or burst suppression pattern EEG. Prognosis in traumatic coma survivors is better than in anoxic cases (Laureys et al. 2009). Recovery from coma may lead to a vegetative state, a minimally conscious state or, more rarely, to a locked-in syndrome (Laureys 2007; Posner et al. 2007; Fig. 2).

1.3 Vegetative State

In 1972, the term "vegetative state" (VS) was defined by Jennet and Plum to describe "an organic body capable of growth and development but devoid of sensation and thought." This state implies the preservation of autonomic functions (e.g., cardio-vascular regulation, thermoregulation) and the sleep–wake cycle with the absence of awareness. Behaviorally, patients in VS open their eyes spontaneously or in response to stimulation, but they only show reflex behaviors, unrelated to the environment (Fig. 2). It is very important to stress the difference between persistent and permanent vegetative states, which are, unfortunately, too often abbreviated identically as PVS, causing unnecessary confusion (Laureys et al. 2000a). When the term "persistent vegetative state" was first described (Jennett and Plum 1972), it was emphasized that persistent did not mean permanent, and it is

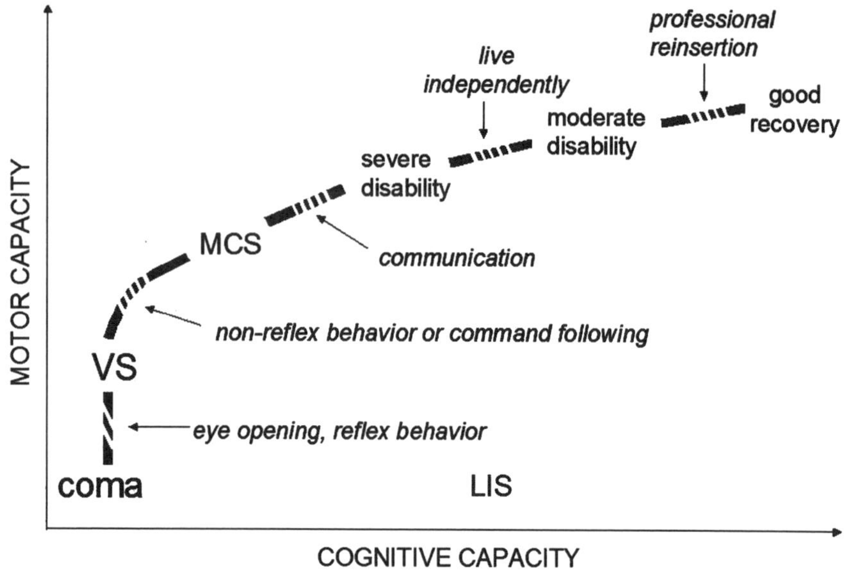

Fig. 2 Different clinical entities encountered in the gradual recovery from coma, illustrated as a function of cognitive and motor capacity. When a patient recovers from a coma to a vegetative state (VS), spontaneous or elicited eye-opening is observed in the absence of voluntary motor activity. The transition from the VS to the minimally conscious state (MCS) is marked by reproducible evidence of "voluntary behavior," defined as (1) non-reflex behavior or simple command following, (2) gestural or verbal yes/no responses (regardless of accuracy), (3) intelligible verbalization, or (4) motor activity occurring in contingent relation to relevant, often emotional, stimuli (also including visual pursuit). Emergence from MCS is signaled by the return of functional communication or object use. The locked-in syndrome (LIS) is the extreme example of intact cognition with nearly complete motor deficit, only permitting eye-coded communication [adapted from Laureys et al. (2005a)]

now recommended that the term "persistent" be avoided and to describe a patient as having been vegetative for a certain period of time. When there is no recovery after a specified period (depending on etiology, 3–12 months), the state can be declared permanent and, only then, can the ethical and legal issues around withdrawal of treatment be discussed (American Congress of Rehabilitation Medicine 1995; Jennett 2005).

1.4 Minimally Conscious State

Patients in a minimally conscious state (MCS) are aroused and show fluctuating but reproducible signs of awareness (Giacino et al. 2002). These patients can manifest oriented behavioral and emotional responses such as response to verbal order, object manipulation, oriented responses to noxious stimulation, visual pursuit or fixation (Fig. 2). The re-emergence of visual pursuit appears to be an early behavioral marker of the transition from VS to MCS (Giacino and Whyte 2005). However, these behaviors can fluctuate in time, which makes the detection of awareness challenging.

Recovery from a MCS is defined by the recuperation of functional communication and/or functional use of objects (Giacino et al. 2002). Nevertheless, some patients can remain in a MCS without fully recovering consciousness for a prolonged period (Fins et al. 2007). The prognosis of recovery in MCS patients remains very difficult because of the huge heterogeneity in recuperation in these patients after a prolonged period of time in a MCS. (Lammi et al. 2005).

1.5 Locked-In Syndrome

Even if it is not an altered state of consciousness, the Locked-in Syndrome (LIS) has to be mentioned as it may present the same behavioral pattern that is observed in VS. Misdiagnosis can therefore easily occur (American Congress of Rehabilitation Medicine 1995; Cairns et al. 1941). Indeed, LIS patients cannot move or talk but are able to use vertical eye movements and blinking to communicate with their surroundings (Fig. 2). This syndrome is often due to a selective supranuclear motor de-efferentation, producing a paralysis of all four limbs and the last cranial nerves without interfering with consciousness (Plum and Posner 1983) or cognition (Schnakers et al. 2005, 2008). According to Bauer et al. (1979), different categories of LIS can be classified on the extent of motor impairment: Classical LIS consists of total immobility but preserved vertical eye movements and blinking; Incomplete LIS is characterized by remnant non-ocular voluntary motions (e.g., head or fingers movements); and Total LIS patients are completely immobile, unable to control any eye movements.

2 Brain Metabolism and Brain Activation Among Vegetative Patients

2.1 Brain Metabolism

In vegetative patients, brain hypometabolism appears similar to individuals in coma, with a 50–60% decrease in global metabolic rate relative to healthy individuals. Hypometabolic activity is further reduced to 60–70% in patients in "permanent" VS (Tommasino et al. 1995). However, the global metabolic rate does not always represent a return to normal after recovery and sometimes does not even represent substantial changes (Laureys et al. 1999a; Fig. 3). Global brain metabolism does not, therefore, reflect the presence or absence of consciousness. Another potential approach consists in comparing brain areas and seeing whether some areas are more important than others. Metabolic dysfunctions were reported in a wide area of the brain, including the polymodal associative cortices: lateral and

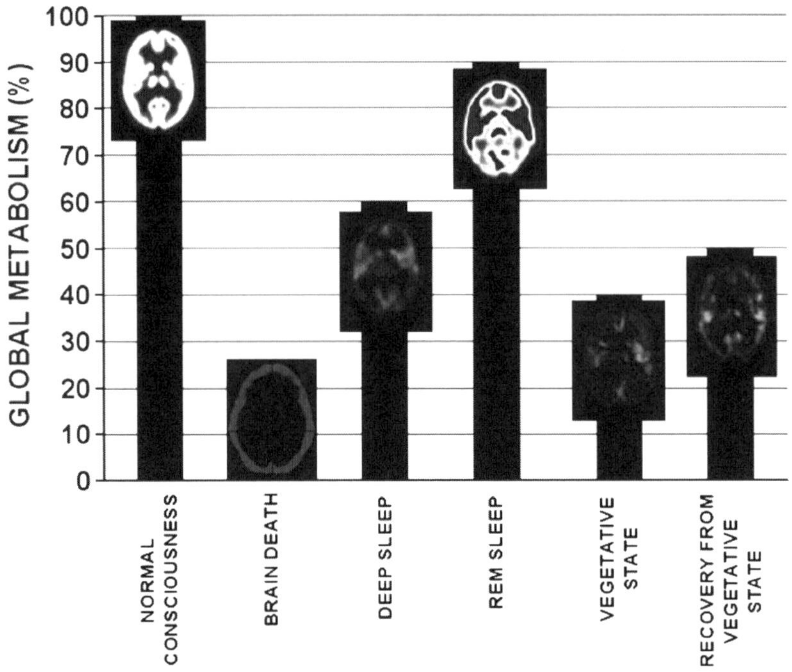

Fig. 3 Cerebral metabolism in conscious healthy subjects and altered states of consciousness. In brain death, cerebral metabolism is absent. In vegetative patients, brain hypometabolism appears similar to individuals in coma, with a 50–60% decrease in global metabolic rate relative to healthy individuals. However, global metabolic rate does not always coming back to normal after recovery and sometimes does not show substantial changes. Global brain metabolism does not, therefore, reflect the presence or absence of consciousness [adapted from Laureys et al. (2004a)]

medial frontal regions bilaterally, parieto-temporal and posterior parietal areas bilaterally, posterior cingulate and precuneal cortices (Laureys et al. 1999b). These impaired areas are involved in the default network, which is the most active in resting, non-stimulated conditions and which is implicated in cognitive processes like daydreaming or mind-wandering, stimulus-independent thoughts, or self-related thoughts (Buckner et al. 2008; Laureys et al. 2007). Despite metabolic dysfunctions in these areas, a recent study showed that the connectivity in the default network was less important in vegetative as compared to conscious patients (i.e., minimally conscious, locked-in and healthy subjects; Vanhaudenhuyse et al. 2010). More specifically, the precuneus (an area known to be crucial in conscious processing; Cavanna and Trimble 2006) was found to be significantly less connected to other default network areas in vegetative patients as compared to minimally conscious patients. In patients having recovered from the vegetative state, there is not only a return to normal metabolism in associative areas but also, and importantly, a re-establishment of the correlation between these areas and the thalamus (Laureys et al. 2000b). The functional connectivity between cortico-cortical areas and thalamo-cortical areas is therefore crucial to recovery of consciousness.

2.2 Brain Activation

The acquisition of spontaneous brain activity can be useful in estimating remnant brain functioning. However, it does not provide information about patients' perception. Previous neuroimaging studies have studied brain processing in response to stimulation. Activation studies using auditory stimulation (i.e., tones) showed preserved functioning in the primary auditory cortex without, nevertheless, encompassing other brain areas (such as the temporoparietal junction; Laureys et al. 2000c). Similarly, noxious stimulation (i.e., electrical stimulation of the median nerve) activated midbrain, contralateral thalamus, and primary somatosensory cortex in vegetative patients without activating brain areas more deeply involved in perception processing, such as secondary somatosensory, bilateral posterior parietal, premotor, polysensory superior temporal, and prefrontal cortices . For both activation studies, low-order primary cortical activity seemed to be isolated from higher-order associative cortical activity (Laureys et al. 2002b; Fig. 4). Finally, a study by Silva et al. (2010) also showed a hypermetabolism in the ascending reticular activating system (ARAS) isolated from the posterior medial cortices (i.e., precuneus) in persistent vegetative patients during proprioceptive stimulation. This residual activity suggests the absence of integration of information and, therefore, the absence of a conscious perception in vegetative patients.

Atypical activation has already been reported (Menon et al. 1998; Di et al. 2007). Nevertheless, most part has afterward evolved towards a MCS suggesting that

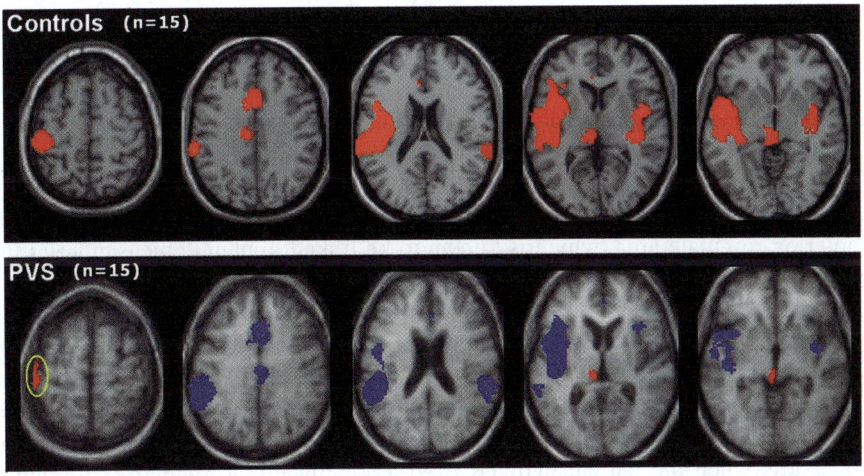

Fig. 4 Cerebral activation to noxious stimulation. Areas in *red* are brain regions that activated the most during noxious stimulation in healthy controls (above) and in permanent vegetative state (PVS, below). In *blue*, brain regions that activated less in PVS than in healthy controls. As highlighted in yellow, in a PVS patient, only the primary areas are activated [adapted from Laureys et al. (2002b)]

neuroimaging techniques could have a prognostic value (Di et al. 2008). Recently, a study of Coleman and co-workers showed that some patients with a diagnosis of VS demonstrated speech comprehension when assessed using functional brain imaging. These fMRI findings were not correlated with the patients' behavioral presentation at the time of investigation but correlated strongly with the patients' recovery 6 months after the scan, suggesting that brain imaging may provide valuable prognostic information. This result has to be replicated to validate neuroimaging techniques as a prognostic tool (Coleman et al. 2009).

3 Differential Diagnosis

3.1 Minimally Conscious State

The metabolic pattern of vegetative patients differs from that of minimally conscious patients. In spite of a total metabolic depression of 20–40% (Schiff et al. 2005), the activity in precuneus and posterior cingulate cortex of minimally conscious patients (the most active regions in awakening and the least active under general anesthesia or during deep slow sleep) is superior to the activity observed in vegetative patients (Laureys et al. 2005a). In contrast with the limited brain activation found in VS, functional imaging studies using auditory stimulation showed a larger temporal activation in minimally conscious patients (encompassing the temporoparietal junction) than in vegetative patients (Bekinschtein et al. 2004; Boly et al. 2004; Schiff et al. 2005). The valence of an auditory stimulation can lead to a difference in brain activation in these patients.

Indeed, minimally conscious patients seem to activate a broader part of the temporal lobe and the amygdala in response to an emotional auditory stimulation than in response to a non-emotional auditory stimulation (Laureys et al. 2004b; Bekinschtein et al. 2004). For other stimulation, such as noxious stimulation, Boly and coworkers showed brain activation similar to controls (involving the anterior cingulate area, which is known to be linked to pain unpleasantness) in minimally conscious patients (Boly et al. 2005, 2008; Fig. 5). These data suggest that these patients may have sufficient cortical integration and access to afferent information to allow conscious perception. Stimuli are therefore processed at a more integrative level. Similarly, a longitudinal case report described by Bekinschtein and co-workers showed a fronto-temporo-parietal activation and no longer an isolated temporal left activation in response to the presentation of words when the patient evolved from a VS towards a MCS (Bekinschtein et al. 2005). In addition to this increased brain activation, a better connectivity between the primary auditory cortex and fronto-temporo-parietal associative cortices was noticed in minimally conscious patients as compared with vegetative patients (Boly et al. 2004, 2008).

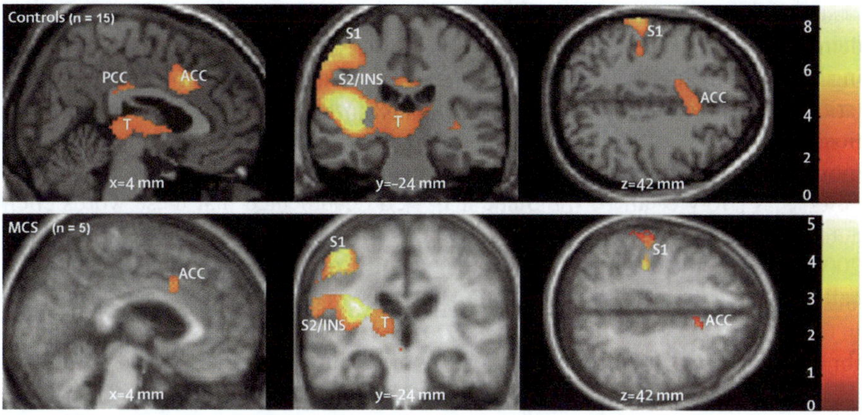

Fig. 5 Response to pain in healthy controls and patients in MCS. On the top, regions of the brain that activated in response to noxious stimulation in controls. On the bottom, brain regions commonly activated during noxious stimulation in MCS patients and in controls. *T* thalamus, *PCC* posterior cingulate cortex, *ACC* anterior cingulate cortex, *S2/INS* secondary somatosensory cortex or insula, *S1* primary somatosensory cortex [adapted from Boly et al. (2008)]

3.2 Locked-In Syndrome

In contrast to vegetative patients, the patients presenting a locked-in syndrome recover a normal consciousness level. As regards the presence of tetraplegia, anarthria but sustained eye opening these patients can easily be misdiagnosed as being in a VS. This misdiagnosis is nevertheless avoidable. Indeed, even if hypometabolism in the cerebellum (linked to the motor deficits) is usually observed, these patients show a preserved brain metabolism (Laureys et al. 2004a). In acute patients, hypermetabolism in the amygdala was also found. As this area is often involved in negative emotions (such as fright), the hypermetabolism observed is probably due to the anxiety of being "blocked" in one's body (Laureys et al. 2005b).

4 Active Paradigms: An Answer to Diagnostic Problems?

Another approach for detecting signs of consciousness consists of explicitly instructing the patients to perform a mental imagery task. This strategy reflects volitional or "consciously willed" neural activation and, therefore, suggests consciousness. This kind of active paradigm seems to be an interesting way to improve diagnosis. We will review different studies showing very interesting results in severely brain-injured patients.

In 2006, Owen et al. reported the case of a young women considered to be clinically in a VS while showing a brain activity indistinguishable from these observed in control healthy subjects when researchers asked her to perform a mental imagery task (Owen et al. 2006). In this patient, activation was observed in the supplementary motor area after she was asked to imagine playing tennis (motor task). In contrast, when asked to imagine visiting her house (spatial navigation task), robust activation was measured in premotor cortex, parahippocampal gyrus and posterior parietal cortex. Similar activation patterns were seen in healthy volunteers (Boly et al. 2007). Since it is well documented that these words elicit automatic neural responses in the absence of conscious awareness (Hauk et al. 2004), some could argue that the words "tennis" and "house" may have automatically triggered the patterns of activation observed in target brain areas in this patient in the absence of conscious awareness. However, such responses typically last only for a few seconds and occur in regions of the brain that are associated with word processing. Here, the observed activity lasted for the full 30 s on each imagery task and persisted until the patient was asked to rest. Moreover, responses were observed in brain regions that are known to be involved in the two imagery tasks that she was asked to carry out (Owen et al. 2007) and not in regions that are known to be involved in word processing. In this sense, the decision to follow the instruction "imagine playing tennis" rather than simply "rest" is an act of willed intention and clearly reflects awareness. It is nevertheless important to underline that this patient was diagnosed as being in a VS according to the UK criteria (Working Party of the Royal College of Physicians 2003) and not according to US criteria (Giacino et al. 2002). Indeed, the patient showed visual fixation, which is considered to be a sign of consciousness by the Aspen Workgroup (Giacino et al. 2002) but not by the Royal College of Physicians (Working Party of the Royal College of Physicians 2003). This patient did not, therefore, present a typical VS. In addition, 6 months later, this patient showed inconsistent signs of awareness, confirming that she was actually already evolving towards a MCS at the time of the study.

Following this study, 54 patients were tested in Liège and Cambridge with the same paradigm. In all of the population tested, only four behaviorally vegetative patients and one minimally conscious patient did the task. We found behavioral signs of consciousness in two of them after re-examination. One of these patients was assessed with a communication protocol, asking him to answer a question with a "yes"/"no" response: by imagining playing tennis to say "yes" and visiting his house to say "no." Interestingly, even if this patient only showed very fluctuating behavioral signs of consciousness, he did the task correctly, suggesting a high level of cognitive functioning (Monti et al. 2010; Fig. 6). This study provides hope by suggesting the possibility of improving the diagnosis and also of finding an alternative way to communicate with severely brain-injured patients, which is of huge importance for ethical issues including treatment and end-of-life questions. Nevertheless, it also shows the importance of being careful with the misinterpretation of the absence of response, as only one minimally conscious patient performed this task. Indeed, these patients often present fluctuating levels of vigilance, which influences task performance.

Recently, an active evoked potentials (ERPs) paradigm was proposed. In this version, the subject was instructed to voluntarily direct his attention and count

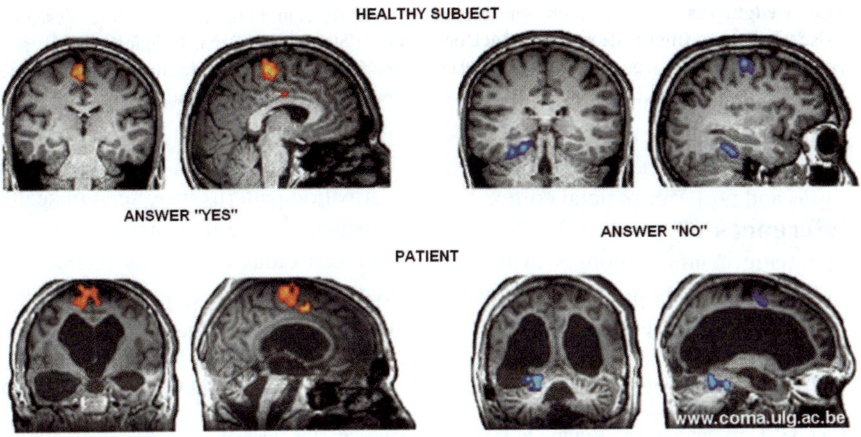

Fig. 6 Willful communication in a patient clinically diagnosed as being in a vegetative state before the study, when he was asked to answer a question with a "yes"/"no" response by imagining playing tennis or visiting his house [adapted from Monti et al. (2010)]

a target stimulus (Schnakers et al. 2008). A larger P3 response (response often linked to complex cognitive functioning such as working memory; Kok 2001) was observed when minimally conscious patients and controls were asked to count a target (e.g., their own name) in a sequence compared to a passive listening condition. The researchers did not find this difference in vegetative patients, suggesting that, in contrast to the minimally conscious patients, vegetative patients were not able to respond to verbal orders, even mentally (Schnakers et al. 2008). However, using the same paradigm, voluntary brain activity was detected in a 21-year-old woman who presented a Total LIS (characterized by tetraplegia, anarthria and paralysis of eye motility) when this patient would behaviorally have been diagnosed as comatose (Schnakers et al. 2009b).

5 Conclusion

Improving the diagnosis of altered states of consciousness is of ethical and medical interest. The alarming rate of misdiagnosis shows how difficult it is to disentangle disorders of consciousness. This review shows that the integration of objective methods such as neuroimaging techniques should improve our diagnosis and the patients' prognosis and should guide our clinical therapeutic options in these challenging patients. Additionally, just as resting metabolism and passive paradigms studies increased our understanding of residual cerebral processing in vegetative and minimally conscious patients, an active paradigm seems to be another powerful objective diagnostic tool to distinguish between these patients.

Acknowledgments This research was funded by the Belgian Fund for Scientific Research (FNRS), European Commission, James McDonnell Foundation, Mind Science Foundation, French Speaking Community Concerted Research Action (ARC-06/11-340), Fondation Médicale Reine Elisabeth, University of Liège and the CNRS/FNRS-CGRI collaboration funds.

References

American Congress of Rehabilitation Medicine (1995) Recommendations for use of uniform nomenclature pertinent to patients with severe alterations of consciousness. Arch Phys Med Rehabil 76:205–209

Andrews K (2004) Medical decision making in the vegetative state: withdrawal of nutrition and hydration. NeuroRehabilitation 19:299–304

Bauer G, Gerstenbrand F, Rumpl E (1979) Varieties of the locked-in syndrome. J Neurol 221:77–91

Bekinschtein T, Leiguarda R, Armony J, Owen A, Carpintiero S, Niklison J, Olmos L, Sigman L, Manes F (2004) Emotion processing in the minimally conscious state. J Neurol Neurosurg Psychiatr 75:788

Bekinschtein T, Tiberti C, Niklison J, Tamashiro M, Ron M, Carpintiero S, Villarreal M, Forcato C, Leiguarda R, Manes F (2005) Assessing level of consciousness and cognitive changes from vegetative state to full recovery. Neuropsychol Rehabil 15:307–322

Boly M, Faymonville ME, Peigneux P, Lambermont B, Damas P, Del Fiore G, Degueldre C, Franck G, Luxen A, Lamy M, Moonen G, Maquet P, Laureys S (2004) Auditory processing in severely brain injured Patients: differences between the minimally conscious state and the persistent vegetative atate. Arch Neurol 61:233–238

Boly M, Faymonville ME, Peigneux P, Lambermont B, Damas F, Luxen A, Lamy M, Moonen G, Maquet P, Laureys S (2005) Cerebral processing of auditory and noxious stimuli in severely brain injured patients: differences between VS and MCS. Neuropsychol Rehabil 15:283–289

Boly M, Coleman MR, Davis MH, Hampshire A, Bor D, Moonen G, Maquet PA, Pickard JD, Laureys S, Owen AM (2007) When thoughts become action: an fMRI paradigm to study volitional brain activity in non-communicative brain injured patients. Neuroimage 36:979–992

Boly M, Faymonville ME, Schnakers C, Peigneux P, Lambermont B, Phillips C, Lancellotti P, Luxen A, Lamy M, Moonen G, Maquet P, Laureys S (2008) Perception of pain in the minimally conscious state with PET activation: an observational study. Lancet Neurol 7:1013–1020

Buckner RL, Andrews-Hanna JR, Schacter DL (2008) The brain's default network: anatomy, function, and relevance to disease. Ann NY Acad Sci 1124:1–38

Cairns H, Oldfield RC, Pennybacker JB, Whitteridge D (1941) Akinetic mutism with an epidermoid cyst at the third ventricle. Brain 64:273–290

Cavanna AE, Trimble MR (2006) The precuneus: a review of its functional anatomy and behavioural correlates. Brain 129:564–583

Coleman MR, Davis MH, Rodd JM, Robson T, Ali A, Owen AM, Pickard JD (2009) Towards the routine use of brain imaging to aid the clinical diagnosis of disorders of consciousness. Brain 132:2541–52

Di HB, Yu SM, Weng XC, Laureys S, Yu D, Li JQ, Qin PM, Zhu YH, Zhang SZ, Chen YZ (2007) Cerebral response to patient's own name in the vegetative and minimally conscious states. Neurology 68:895–899

Di H, Boly M, Weng X, Ledoux D, Laureys S (2008) Neuroimaging activation studies in the vegetative state: predictors of recovery? Clin Med 8:502–507

Fins JJ, Schiff ND, Foley KM (2007) Late recovery from the minimally conscious state: ethical and policy implications. Neurology 68:304–307

Giacino J, Whyte J (2005) The vegetative and minimally conscious states: current knowledge and remaining questions. J Head Trauma Rehabil 20:30–50
Giacino J, Ashwal S, Childs N, Cranford R, Jennett B, Katz D, Kelly J, Rosenberg J, Whyte J, Zafonte R (2002) The minimally conscious state: definition and diagnostic criteria. Neurology 58:349–353
Gill-Thwaites H (2006) Lotteries, loopholes and luck: misdiagnosis in the vegetative state patient. Brain Inj 20:1321–1328
Guidelines for the determination of death (1981) Report of the medical consultants on the diagnosis of death to the President's Commission for the study of ethical problems in medicine and biomedical and behavioral research. J Am Med Assoc 246:2184–2186
Hauk O, Johnsrude I, Pulvermuller F (2004) Somatotopic representation of action words in human motor and premotor cortex. Neuron 41:301–307
Jennett B (2005) The assessment and rehabilitation of vegetative and minimally conscious patients: definitions, diagnosis, prevalence and ethics. Neuropsychol Rehabil 15:163–165
Jennett B, Plum F (1972) Persistent vegetative state after brain damage. A syndrome in search of a name. Lancet 1:734–737
Kok A (2001) On the utility of P3 amplitude as a measure of processing capacity. Psychophysiology 38:557–577
Lammi MH, Smith VH, Tate RL, Taylor CM (2005) The minimally conscious state and recovery potential: a follow-up study 2 to 5 years after traumatic brain injury. Arch Phys Med Rehabil 86:746–754
Laureys S (2007) Eyes open, brain shut: the vegetative state. Sci Am 4:32–37
Laureys S, Lemaire C, Maquet P, Phillips C, Franck G (1999a) Cerebral metabolism during vegetative state and after recovery to consciousness. J Neurol Neurosurg Psychiatr 67:121
Laureys S, Goldman S, Phillips C, Van Bogaert P, Aerts J, Luxen A, Franck G, Maquet P (1999b) Impaired effective cortical connectivity in vegetative state: preliminary investigation using PET. Neuroimage 9:377–382
Laureys S, Faymonville ME, Degueldre C, Fiore GD, Damas P, Lambermont B, Janssens N, Aerts J, Franck G, Luxen A, Moonen G, Lamy M, Maquet P (2000a) Auditory processing in the vegetative state. Brain 123:1589–15601
Laureys S, Faymonville ME, Luxen A, Lamy M, Franck G, Maquet P (2000b) Restoration of thalamocortical connectivity after recovery from persistent vegetative state. Lancet 355:1790–1791
Laureys S, Faymonville ME, Berre J (2000c) Permanent vegetative state and persistent vegetative state are not interchangeable terms. In: rapid response to Dyer C (2000) October 17, online
Laureys S, Faymonville M, Maquet P (2002a) Quelle conscience durant le coma? Pour la Science 302:122–128
Laureys S, Faymonville ME, Peigneux P, Damas P, Lambermont B, Del Fiore G, Degueldre C, Aerts J, Luxen A, Franck G (2002b) Cortical processing of noxious somatosensory stimuli in the persistent vegetative state. Neuroimage 17:732–741
Laureys S, Owen A, Schiff D (2004a) Brain function in coma, vegetative state and related disorders. Lancet Neurol 3:537–546
Laureys S, Perrin F, Faymonville M, Schnakers C, Boly M, Bartsch V, Majerus S, Moonen G, Maquet P (2004b) Cerebral processing in the minimally conscious state. Neurology 63:916–918
Laureys S, Perrin F, Schnakers C, Boly M, Majerus S (2005a) Residual cognitive function in comatose, vegetative and minimally conscious states. Curr Opin Neurol 18:726–733
Laureys S, Pellas F, Van Eeckhout P, Ghorbel S, Schnakers C, Perrin F, Berre J, Faymonville ME, Pantke KH, Damas F, Lamy M, Moonen G, Goldman S (2005b) The locked-in syndrome: what is it like to be conscious but paralyzed and voiceless? Prog Brain Res 150:495–511
Laureys S, Perrin F, Bredart S (2007) Self-consciousness in non-communicative patients. Conscious Cogn 16:722–741

Laureys S, Boly M, Moonen G, Maquet P (2009) Coma. Encyclopedia of Neuroscience. In: Squire L (ed) Academic Press, Oxford UK pp: 1133–1142
Majerus S, Gill-Thwaites H, Andrews K, Laureys S (2005) Behavioral evaluation of consciousness in severe brain damage. Prog Brain Res 150:397–413
Menon DK, Owen AM, Williams EJ, Minhas PS, Allen CMC, Boniface SJ, Pickard JD (1998) Cortical processing in persistent vegetative state. Lancet 352:200
Monti MM, Vanhaudenhuyse A, Coleman MR, Boly M, Pickard J, Tshibanda J-F, Owen AM, Laureys S (2010) Willful modulation of brain activity in disorders of consciousness. New Engl J Med 362:579–589
Owen AM, Coleman MR, Boly M, Davis MH, Laureys S, Pickard JD (2006) Detecting awareness in the vegetative state. Science 313:1402
Owen AM, Coleman MR, Boly M, Davis MH, Laureys S, Pickard JD (2007) Using functional magnetic resonance imaging to detect covert awareness in the vegetative state. Arch Neurol 64:1098–1102
Plum F, Posner JB (1983) The diagnosis of stupor and coma. Davis FA, Philadelphia, PA
Posner JB, Saper CB, Plum F (2007) Plum and posner's diagnosis of stupor and coma. In: Davis (ed) The diagnosis of stupor and coma. Oxford University Press, New York
Schiff ND, Rodriguez-Moreno D, Kamal A, Kim KH, Giacino JT, Plum F, Hirsch J (2005) fMRI reveals large-scale network activation in minimally conscious patients. Neurology 64:514–523
Schnakers C, Majerus S, Laureys S, Van Eeckhout P, Peigneux P, Goldman S (2005) Neuropsychological testing in chronic locked-in syndrome. Psyche, abstracts from the Eighth Conference of the Association for the Scientific Study of Consciousness (ASSC8). University of Antwerp, Belgium, p 11
Schnakers C, Perrin F, Schabus M, Majerus S, Ledoux D, Damas P, Boly M, Vanhaudenhuyse A, Bruno MA, Moonen G, Laureys S (2008) Voluntary brain processing in disorders of consciousness. Neurology 71:1614–1620
Schnakers C, Vanhaudenhuyse A, Giacino J, Ventura M, Boly M, Majerus S, Moonen G, Laureys S (2009a) Diagnostic accuracy of the vegetative and minimally conscious state: clinical consensus versus standardized neurobehavioral assessment. BMC Neurol 9:35
Schnakers C, Perrin F, Schabus M, Hustinx R, Majerus S, Moonen G, Boly M, Vanhaudenhuyse A, Bruno MA, Laureys S (2009b) Detecting consciousness in a total locked-in syndrome: an active event-related paradigm. Neurocase 15:1–7
Silva S, Alacoque X, Fourcade O, Samii K, Marque P, Woods R, Mazziotta J, Chollet F, Loubinoux I (2010) Wakefulness and loss of awareness: brain and brainstem interaction in the vegetative state. Neurology 74:313–320
Tommasino C, Grana C, Lucignani G, Torri G, Fazio F (1995) Regional cerebral metabolism of glucose in comatose and vegetative state patients. J Neurosurg Anesthesiol 7:109–116
Vanhaudenhuyse A, Noirhomme Q, Tshibanda LJ, Bruno MA, Boveroux P, Schnakers C, Soddu A, Perlbarg V, Ledoux D, Brichant JF, Moonen G, Maquet P, Greicius MD, Laureys S, Boly M (2010) Default network connectivity reflects the level of consciousness in non-communicative brain-damaged patients. Brain 133:161–171
Working Party of the Royal College of Physicians (2003) The vegetative state: guidance on diagnosis and management. Clin Med 3:249–254

When Thoughts Become Actions: Imaging Disorders of Consciousness

Adrian M. Owen

Abstract How can we ever know, unequivocally, that another person is aware? Notwithstanding deeper philosophical considerations about the nature of consciousness itself, the only reliable method we have for detecting awareness in others is by eliciting a predicted response to an external prompt or command. Logically, therefore, our ability to detect awareness in others is determined not by whether they are aware or not but by their ability to communicate that fact through a recognised behavioural response. This problem exposes a central conundrum in the study of awareness in general and, in particular, how it relates to the vegetative state. From this perspective, I discuss various solutions to this problem using functional neuroimaging. In particular, I will contrast those circumstances in which fMRI data can be used to infer awareness in the absence of a reliable behavioural response with those circumstances in which it cannot.

1 Introduction

How can any of us be absolutely sure that another human is consciously aware? When I ask this question, I am not seeking to raise any deep philosophical notions about the nature of consciousness itself but rather to pose a much more pragmatic, down-to-earth question: How can I know that you (or any other person) is aware? By this I mean, how can I know that you are aware of who you are, aware of where you are (in time and space) and aware of what you are doing right now (reading this article)? The answer is that I can only really know if you tell me, via some form of recognised behavioural response. That response may be a spoken answer or a non-verbal signal (which may be a movement as simple as the blink of an eye), but it is that response, and only that response, that would allow me to infer awareness.

A.M. Owen (✉)
Centre for Brain and Mind, Room 224, Natural Sciences Building, The University of Western Ontario, London, Ontario, N6A 5B7, Canada
e-mail: adrian.owen@uwo.ca

S. Dehaene and Y. Christen (eds.), *Characterizing Consciousness: From Cognition to the Clinic?* Research and Perspectives in Neurosciences,
DOI 10.1007/978-3-642-18015-6_6, © Springer-Verlag Berlin Heidelberg 2011

Thus, *awareness* is an internal state of being that can only be measured via some form of self-report. In this sense, it differs fundamentally from that other central pillar of consciousness, wakefulness, which can be measured and monitored accurately by simple observation (if your eyes are open, then it is very likely that you are 'awake') or by using techniques such as electroencephalography (EEG) to measure the electrical signals that characterise the normal waking state. Thus, our ability to know unequivocally that another being is consciously *aware* is determined not by whether they are aware or not but by their ability to communicate that fact through a recognised behavioural response. But what if the ability to speak, blink an eye or move a hand is lost, yet conscious awareness remains?

In recent years, improvements in intensive care have lead to an increase in the number of patients who survive severe brain injury. Although some of these patients go on to make a good recovery, many do not, and some of these individuals progress to a condition known as the vegetative state. Central to the description of this complex condition is the concept of 'wakefulness without awareness,' according to which vegetative patients are assumed to be entirely unaware, despite showing clear signs of wakefulness (Jennett and Plum 1972). However, the assessment of these patients is extremely difficult and relies heavily on subjective interpretation of observed behaviour at rest and in response to stimulation. A diagnosis is made after repeated examinations have yielded no evidence of sustained, reproducible, purposeful or voluntary behavioural response to visual, auditory, tactile or noxious stimuli. Thus, a positive diagnosis (of vegetative state) is ultimately dependent on a negative finding (no signs of awareness) and is, therefore, inherently vulnerable to a Type II error or a false negative result. Indeed, internationally agreed diagnostic criteria for the vegetative state repeatedly emphasize the notion of 'no evidence of awareness of environment or self' – in this instance, absence of evidence does appear to be considered adequate evidence of absence.

Indeed, any assessment that is based on exhibited behaviour after brain injury will be prone to error for a number of reasons. First, an inability to move and speak is a frequent outcome of chronic brain injury and does not necessarily imply a lack of awareness. Second, the behavioural assessment is highly subjective: behaviours such as smiling and crying are typically reflexive and automatic, but in certain contexts they may be the only means of communication available to a patient and therefore reflect a wilful, volitional act of intention. These difficulties, coupled with inadequate experience and knowledge engendered through the relative rarity of these complex conditions, contribute to an alarmingly high rate of misdiagnosis (up to 43%) in this patient group (Andrews et al. 1996; Childs et al. 1993; Schnakers et al. 2006).

These issues expose a central conundrum in the study of covert awareness in general and, in particular, how it relates to conditions such as the vegetative state. Following the logic above, in a case where every opportunity for self-report has been lost (in fact, this is a central requirement for a diagnosis of vegetative state), it would be impossible to determine whether any level of awareness remains. Of course, cases of locked-in syndrome following acute brain injury or disease have been reported for many years, but where such cases are unexpectedly identified it is always through the (sometimes chance) detection of a minor residual motor response. Against this background it is an unfortunate, but inevitable, fact that a population of patients will exist who retain at least some level of residual conscious awareness, yet remain entirely unable to convey that fact to those around them.

Recent advances in neuroimaging technology may provide a solution to this problem. If measurable brain responses could be marshalled and used as a proxy for a motor response, then a patient who is entirely unable to move may be able to signal awareness by generating a pattern of brain activity that is indicative of a specific thought or intention. In this chapter, those circumstances in which fMRI data can be used to infer awareness in the absence of a behavioural response will be contrasted with those circumstances in which it cannot. This distinction is fundamental for understanding and interpreting patterns of brain activity following acute brain injury and has implications for clinical care, diagnosis, prognosis and medical–legal decision-making after serious brain injury.

2 An Historical Perspective

Substantial evidence now exists to suggest that so-called 'activation' methods, such as $H_2{}^{15}O$ positron emission tomography (PET) and fMRI, can be used to link changes in regional cerebral blood flow to specific cognitive processes without the need for any overt response (e.g., a motor action or a verbal response; for review, see Owen et al. 2001). In the first study of its kind, de Jong et al. (1997) measured regional cerebral blood flow in a post-traumatic vegetative patient during an auditorily presented story told by his mother. Compared to non-word sounds, activation was observed in the anterior cingulate and temporal cortices, possibly reflecting emotional processing of the contents, or tone, of the mother's speech. A year later, PET was used in another patient diagnosed as vegetative to study visual processing in response to familiar faces (Menon et al. 1998). Robust activity was observed in the right fusiform gyrus, the so-called human face area (or FFA). In both of these early cases, normal brain activation was observed in the absence of any behavioural responses to the external sensory stimulation.

More recently, in the largest study to date, 41 patients with disorders of consciousness were graded according to their brain activation on a hierarchical series of language paradigms (Coleman et al. 2009). The tasks increased in complexity systematically from basic acoustic processing (a non-specific response to sound) to more complex aspects of language comprehension and semantics. At the highest level, responses to sentences containing semantically ambiguous words (e.g., the *creak/creek* came from a *beam* in the *ceiling/sealing*) are compared to sentences containing no ambiguous words (e.g., her secrets were written in her diary) in order to reveal brain activity associated with spoken language comprehension (Rodd et al. 2005; Owen et al. 2002, 2005a, b; Coleman et al. 2007, 2009). Nineteen of the patients (almost 50%) who had been diagnosed as either vegetative or minimally conscious showed normal or near normal temporal-lobe responses in the low-level auditory contrast (sound responses) and in the mid-level speech perception contrast (a specific response to speech over and above the more general response to sounds). Four patients, including two who had been diagnosed as behaviorally vegetative, were also shown to exhibit normal fMRI activity during the highest-level speech comprehension task, suggesting that the neural processes involved in understanding speech were also intact (Coleman et al. 2009). What is most remarkable about these fMRI findings is that the imaging results were found to have no association with the patients' behavioural presentation at the time of investigation and thus provide additional diagnostic information beyond the traditional clinical assessment. Moreover, the level of auditory processing revealed by the fMRI results did correlate strongly with the patients' subsequent behavioural recovery (assessed 6 months after the scan), suggesting that brain imaging may also provide valuable prognostic information not evident through bedside testing. These results provide compelling evidence for intact, high-level residual linguistic processing in some patients who behaviourally meet the clinical criteria for vegetative and minimally conscious states.

3 On the Relationship Between Brain Activity and Awareness

Does the presence of normal brain activation in behaviourally non-responsive patients indicate awareness? In most of the cases discussed above and elsewhere in the literature, the answer to this question is probably "no." Many types of stimuli, including faces, speech and pain, will elicit relatively automatic responses from the brain; that is to say, they will occur without the need for active (i.e., conscious) intervention on the part of the participant (e.g., you cannot choose to not recognise a face or to not understand speech that is presented clearly in your native language). In addition, a wealth of data in healthy volunteers, from studies of implicit learning and the effects of priming (see Schacter 1994 for review) to studies of learning and speech perception during anaesthesia (e.g., Davis et al. 2007; Bonebakker et al. 1996), have demonstrated that many aspects of human cognition can go on in the absence of awareness. Even the semantic content of masked information can be primed to affect subsequent behaviour without the explicit knowledge of the participant, suggesting that some aspects of semantic processing may occur without conscious awareness (Dehaene et al. 1998). By the same argument, normal neural responses in patients who are diagnosed as vegetative do not necessarily indicate that these patients have any conscious experience associated with processing those same types of stimuli. To investigate this issue directly, Davis et al. (2007) recently used fMRI in sedated healthy volunteers and exposed them to exactly the same speech stimuli that have been shown to elicit normal patterns of brain activity in some vegetative and minimally conscious patients (Owen et al. 2005a, b; Coleman et al. 2007, 2009). During three scanning sessions, the participants were non-sedated (awake), lightly sedated (a slowed response to conversation) and deeply sedated (no conversational response, rousable by loud command). In each session, they were exposed to sentences containing ambiguous words, matched sentences without ambiguous words and signal-correlated noise. Equivalent temporal-lobe responses for normal speech sentences compared to signal-correlated noise were observed, bilaterally, at all three levels of sedation, suggesting that a normal brain response to speech sounds is not a reliable correlate of awareness. This result suggests that extreme caution needs to be exercised when interpreting normal responses to speech in patients who are diagnosed as vegetative, a problem of interpretation that applies to many of the activation studies that have been conducted in vegetative patients to date.

However, when Davis et al. (2007) examined the effects of anaesthesia on ambiguous sentences, the frontal-lobe and posterior temporal-lobe activity that occurs in the awake individual (and is assumed to be a neural marker for semantic processing) was markedly absent, even during light sedation. This finding suggests that vegetative patients who show this specific pattern of neural activity during the presentation of ambiguous semantic material may be consciously aware (e.g., Owen et al. 2005a, b; Coleman et al. 2007, 2009). However, as tantalizing as such conclusions might be, they are entirely speculative; the fact that awareness is associated with the activity changes that are thought to reflect sentence comprehension does not mean that it is necessary for them to occur (by simple analogy, the fact that amygdala activity is often observed during fMRI studies of fear does not mean that, in all studies that have reported amygdala activity, the participants were fearful).

4 Brain Activity as a Form of Response

The studies described above confirm that many of the brain responses that have been observed to date using fMRI in brain damaged patients could have occurred automatically, that is, they could have occurred in the absence of any awareness of self (or others) on the part of the patient. But let us now consider an entirely different type of brain imaging experiment in which the responses observed cannot occur in the absence of awareness, because they are necessarily guided by a conscious choice, or decision, on the part of the participant. When healthy participants in the scanner are asked to imagine moving their arms, robust activity is observed in the premotor cortex (Owen et al. 2006; Boly et al. 2007; Owen and Coleman 2008b; Monti et al. 2010). This response is utterly reliable, even at the single-subject level, and can be observed when the motor imagery task is compared to periods of rest or to other types of imagery tasks, such as navigating one's way around a familiar environment (Owen et al. 2006; Boly et al. 2007). Indeed, the localised changes in fMRI signal associated with these mental activities are so reliable that they can be used in place of a more traditional behavioural (e.g., motor) response, that is, as a proxy for a motor action or what I shall henceforth call a 'brain act.' For example, when healthy participants are asked to answer simple "yes/no" questions in the scanner (e.g., "Have you ever been to Paris?") by imagining moving their arms (and thereby activating the premotor cortex) to convey the answer "yes" and to relax (and thereby not activating the premotor cortex) to convey the answer "no," the factually correct answer can be decoded from their brain activity with 100% accuracy (Owen and Coleman 2008b; see also Monti et al. 2010). Importantly, this approach differs from all of the passive tasks described above (e.g., speech or face perception) because the pattern of fMRI activity is entirely dependent on the participant making a conscious choice to exert a specific wilful, or voluntary, response, rather than the stimulus per se. For example, an entirely different pattern of fMRI activity (e.g., either an increase in premotor cortex indicating a "yes" response or a lack of increase indicating a "no" response) may be generated following an identical physical stimulus ("Have you ever been to Paris?"), depending on whether the factually correct answer is "yes" or "no."

Thus, like any other form of action that requires response selection, such brain acts require awareness of the various contingencies that govern the relationship between any given stimulus (in this case, a question that has a factual answer) and a response (in this case, one of two possibilities that will only be known when the factually correct answer has been accessed from long-term memory). Put simply, fMRI responses of this sort can be used to measure awareness because awareness is necessary for them to occur. Indeed, the fact that they occur at all allows an observer to conclude not only that the instigator of the response is aware but also that multiple cognitive processes that are typically associated with conscious awareness are also intact and working normally. For example, an intact long-term memory is required to access the factually correct answer, short-term (or 'working') memory is required to maintain attention between the stimulus and the response and to guide the search for the correct answer, attentional switching is required (as the instigator of the responses switches between the various mental states that code for "yes" and "no"), sustained attention is required to maintain the appropriate mental state (typically for 30 s at a time) and, of course, response selection is required to make the final decision about which brain act to initiate. In short, because brain acts represent a neural proxy for motor behaviour, they also confirm that the participant retains the ability to understand instructions, to carry out different mental tasks in response to those instructions and, therefore, is able to exhibit willed, voluntary behaviour in the absence of any overt action. On this basis, they permit the identification of awareness at the single-subject level, without the need for a motor response (for discussion, see Owen and Coleman 2008a; Monti et al. 2009).

This contrast between the responses observed in passive fMRI tasks that are (or at least could be) elicited automatically by an external stimulus and active tasks in which the response itself (the brain act) represents a conscious choice is absolutely central to the debate about the use of functional neuroimaging to measure covert awareness. A significant development in this field, therefore, has been application of such paradigms in patients who are entirely behaviourally non-responsive (Owen et al. 2006; Boly et al. 2007; Owen and Coleman 2008b; Monti et al. 2010). In one recent study (Boly et al. 2007), 34 healthy volunteers were asked to imagine hitting a tennis ball back and forth to an imaginary coach when they heard the word 'tennis' (thereby eliciting vigorous imaginary arm movements) and to imagine walking from room to room in their house when they heard the word 'house' (thereby eliciting imaginary spatial navigation). Imagining playing tennis was associated with robust activity in the supplementary motor area in each and every one of the participants scanned. In contrast, imagining moving from room to room in a house activated the parahippocampal cortices, the posterior parietal lobe and the lateral premotor cortices, all regions that have been shown to contribute to imaginary, or real, spatial navigation (Aguirre et al. 1996; Boly et al. 2007).

In severe brain injury, when the request to move a hand or a finger is followed by an appropriate motor response, the diagnosis can change from vegetative state (no evidence of awareness) to minimally conscious state (some evidence of awareness). By analogy then, if the request to activate, say, the supplementary motor area of the brain by imagining moving the hand was followed by an appropriate brain response, shouldn't we give that response the very same weight? Sceptics may argue that brain responses are somehow less physical, reliable or immediate than motor responses but, as is the case with motor responses, all of these arguments can be dispelled with careful measurement, replication and objective verification. For example, if a patient who was assumed to be unaware raised his/her hand to command on just one occasion, there would remain some doubt about the presence of awareness given the possibility that this movement was a chance occurrence, coincident with the instruction. However, if that same patient were able to repeat this response to command on ten occasions, there would remain little doubt that the patient was aware. By the same token, if that patient was able to activate his/her supplementary motor area in response to command (e.g., by being told to imagine hand movements), and was able to do this on every one of ten trials, would we not have to accept that this patient was consciously aware?

This same logic was used recently to demonstrate that a young woman who fulfilled all internationally agreed criteria for the vegetative state was, in fact, consciously aware and able to make responses of this sort using her brain activity (Owen et al. 2006, 2007). Prior to the fMRI scan, the patient was instructed to perform the two mental imagery tasks described above. When she was asked to imagine playing tennis, significant activity was observed in the supplementary motor area (Owen et al. 2006) that was indistinguishable from that observed in the healthy volunteers scanned by Boly et al. (2007). Moreover, when she was asked to imagine walking through her home, significant activity was observed in the parahippocampal gyrus, the posterior parietal cortex and the lateral premotor cortex, which was again indistinguishable from the activity observed in healthy volunteers (Owen et al. 2006, 2007). On this basis, it was concluded that, despite fulfilling all of the clinical criteria for a diagnosis of vegetative state, this patient retained the ability to understand spoken commands and to respond to them through her brain activity rather than through speech or movement, confirming beyond any doubt that she was consciously aware of herself and her surroundings. In a follow-up study of 23 patients who were behaviourally diagnosed as vegetative, Monti et al. (2010) showed that four (17%) were able to generate reliable responses of this sort in the fMRI scanner.

Owen and Coleman (2008b) extended the general principle described above, by which active mental rehearsal is used to signify awareness, to show that communication of "yes" and "no" responses is possible using the same approach. Thus, a healthy volunteer was able to reliably convey a "yes" response by imagining playing tennis and a "no" response by imagining moving around a house, thereby providing the answers to simple questions posed by the experimenters using only his brain activity. This technique was further refined by Monti et al. (2010), who successfully decoded the "yes" and "no" responses of 16 healthy participants with 100% accuracy using only their real time changes in the supplementary motor area (during tennis imagery) and the parahippocampal place area (during spatial navigation). Moreover, in one traumatic brain injury patient who had been repeatedly diagnosed as vegetative over a 5-year period, similar questions were posed and successfully decoded using the same approach (Monti et al. 2010). However, despite a re-classification to minimally conscious state following the fMRI scan, it remained impossible to establish any form of communication with this patient at the bedside.

Clearly, the patient described by Monti et al. (2010) was not vegetative because he could generate "yes" and "no" responses in real time by wilfully modulating his brain activity. In fact, these consistent 'responses to command' which allowed him to functionally communicate suggest a level of residual cognitive function that would actually place this patient beyond the minimally conscious state and (at least) into the severely disabled category. Likewise, the patient described by Owen et al. (2006) was clearly able to produce voluntary responses to command (albeit neural responses) yet was unable to match this with any form of motor response at the bedside. Paradoxically, therefore, her (motor) behaviour was consistent with a diagnosis of vegetative state (an absence of evidence of awareness or purposeful response) yet her brain imaging data confirmed that the alternative hypothesis was correct, i.e., that she was entirely aware during the scanning procedure.

These types of approaches all illustrate a paradigmatic shift away from passive (e.g., perceptual) tasks to more active (e.g., wilful) tasks in the assessment of covert awareness after serious brain injury. What sets such tasks apart is that the neural responses required are not produced automatically by the eliciting stimulus but rather depend on time-dependent and sustained responses generated by the participant. Such behaviour (albeit neural behaviour) provides a proxy for a motor action and is, therefore, an appropriate vehicle for reportable awareness (Zeman 2009).

Of course, sceptics may still argue that brain acts do not 'prove' that a person is consciously aware, even in situations where behaviourally unresponsive patients have been able to use this method to provide factually correct answers to five biographical questions about themselves (Monti et al. 2010). However, such sceptics would likely remain unsatisfied even if 500 questions had been asked and 500 correct answers had been given. The important point is that, by using spatially and temporally reliable fMRI changes as willed responses, we are simply applying the very same criteria that any of us would have to use to determine that any other walking, talking human being was consciously aware, that is, we would pose a question (or a series of questions) and we would make our judgement based on the response(s). It is difficult to imagine a circumstance in which any of us would spontaneously ask another person a series of five questions (drawn from an almost limitless pool of possible questions that could be asked), receive five factually correct answers and then conclude that they were not consciously aware (NB: Philosophically of course, it is possible to imagine that such a person could exist, wholly unaware, yet able to answer an infinite number of questions with factually correct answers, yet in the absence of any data to suggest that such a person does or can exist, I will not consider this possibility any further). Of course, it is true that it is impossible to know much about a patient's internal mental world on the basis of the answers to five simple questions, but the important point is that, as long as an answer can be conveyed with a "yes" or a "no," the possibility now exists for them to be asked. Indeed, there is no reason why such a patient could not be asked (and could answer) the most difficult question of all: "Are you conscious?"

Acknowledgements I would like to thank the James S. McDonnell Foundation, the Medical Research Council, U.K. and the Canada Excellence Research Chairs Programme for their generous funding of my research programme.

References

Aguirre GK, Detre JA, Alsop DC, D'Esposito M (1996) The parahippocampus subserves topographical learning in man. Cereb Cortex 6:823–829

Andrews K, Murphy L, Munday R, Littlewood C (1996) Misdiagnosis of the vegetative state: retrospective study in a rehabilitation unit. Br Med J 313:13–16

Boly M, Coleman MR, Davis MH, Hampshire A, Bor D, Moonen G, Maquet PA, Pickard JD, Laureys S, Owen AM (2007) When thoughts become action: an fMRI paradigm to study volitional brain activity in non-communicative brain injured patients. Neuroimage 36:979–992

Bonebakker A, Bonke B, Klein J, Wolters G, Stijnen T, Passchier J, Merikle PM (1996) Information processing during general anaesthesia: evidence for unconscious memory. In: Bonke B, Bovill JGW, Moerman N (eds) Memory and awareness in anaesthesia. Swets and Zeitlinger, Lisse, Amsterdam, pp 101–109

Childs NL, Mercer WN, Childs HW (1993) Accuracy of diagnosis of persistent vegetative state. Neurology 43:1465–1467

Coleman MR, Rodd JM, Davis MH, Johnsrude IS, Menon DK, Pickard JD, Owen AM (2007) Do vegetative patients retain aspects of language: evidence from fMRI. Brain 130:2494–2507

Coleman MR, Davis MH, Rodd JM, Robson T, Ali A, Pickard JD, Owen AM (2009) Towards the routine use of brain imaging to aid the clinical diagnosis of disorders of consciousness. Brain 132:2541–2552

Davis MH, Coleman MR, Absalom AR, Rodd JM, Johnsrude IS, Matta BF, Owen AM, Menon DK (2007) Dissociating speech perception and comprehension at reduced levels of awareness. Proc Natl Acad Sci USA 104:16032–16037

de Jong B, Willemsen AT, Paans AM (1997) Regional cerebral blood flow changes related to affective speech presentation in persistent vegetative state. Clin Neurol Neurosurg 99:213–216

Dehaene S, Naccache L, Le Clec'H G, Koechlin E, Mueller M, Dehaene-Lambertz G, Van De Moortele PF, Le Bihan D (1998) Imaging unconscious semantic priming. Nature 395:597–600

Jennett B, Plum F (1972) Persistent vegetative state after brain damage. Lancet 1:734–737

Menon DK, Owen AM, Williams EJ, Minhas PS, Allen CMC, Boniface S, Pickard JD (1998) Cortical processing in persistent vegetative state. Lancet 352:200

Monti MM, Coleman MR, Owen AM (2009) Executive functions in the absence of behavior: functional imaging of the minimally conscious state. In: Laureys S, Schiff ND, Owen AM (eds) Coma science: clinical and ethical implications. Progress in brain research. Oxford, Elsevier, pp 249–260

Monti MM, Vanhaudenhuyse A, Coleman MR, Boly M, Pickard JD, Tshibanda JFL, Owen AM, Laureys S (2010) Willful modulation of brain activity and communication in disorders of consciousness. New Engl J Med 362:579–589

Owen AM, Coleman M (2008a) Functional imaging in the vegetative state. Nat Rev Neurosci 9:235–243

Owen AM, Coleman MR (2008b) Detecting awareness in the vegetative state. In: Pfaff D (ed) Molecular and biophysical mechanisms of arousal, alertness and attention. Ann NY Acad Sci 1129:130–140

Owen AM, Epstein R, Johnsrude IS (2001) fMRI: Applications to Cognitive Neuroscience. In: Jezzard P, Mathews PM, Smith SM (eds) Functional Magnetic Resonance Imaging. An Introduction to Methods. Oxford University Press, Oxford, UK

Owen AM, Menon DK, Johnsrude IS, Bor D, Scott SK, Manly T, Williams EJ, Mummery C, Pickard JD (2002) Detecting residual cognitive function in persistent vegetative state. Neurocase 8:394–403

Owen AM, Coleman MR, Menon DK, Berry EL, Johnsrude IS, Rodd JM, Davis MH, Pickard JD (2005a) Using a heirarchical approach to investigate residual auditory cognition in persistent vegetative state. In: Laureys S (ed) The boundaries of consciousness: neurobiology and neuropathology, vol 150, Progress in brain research. Elsevier, London, pp 461–476

Owen AM, Coleman MR, Menon DK, Johnsrude IS, Rodd JM, Davis MH, Taylor K, Pickard JD (2005b) Residual auditory function in persistent vegetative state: a combined PET and fMRI study. Neuropsychol Rehabil 15:290–306

Owen AM, Coleman MR, Davis MH, Boly M, Laureys S, Pickard JD (2006) Detecting awareness in the vegetative state. Science 313:1402

Owen AM, Coleman MR, Davis MH, Boly M, Laureys S, Jolles D, Pickard JD (2007) Response to comments on "Detecting awareness in the vegetative state". Science 315:1221c

Rodd JM, Davis MH, Johnsrude IS (2005) The neural mechanisms of speech comprehension: fMRI studies of semantic ambiguity. Cereb Cortex 15:1261–1269

Schacter DL (1994) Priming and multiple memory systems: perceptual mechanisms of implicit memory. In: Schacter DL, Tulving E (eds) Memory systems. MIT, Cambridge, MA, pp 233–268

Schnakers C, Giacino J, Kalmar K, Piret S, Lopez E, Boly M, Malone R, Laureys S (2006) Does the FOUR score correctly diagnose the vegetative and minimally conscious states? Ann Neurol 60:744–745

Zeman A (2009) The problem of unreportable awareness. In: Laureys S, Schiff ND, Owen AM (eds) Coma science: clinical and ethical implications, Progress in brain research. Oxford, Elsevier, pp 1–10

Rhythmic Neuronal Synchronization Subserves Selective Attentional Processing

Thilo Womelsdorf and Pascal Fries

Abstract Selective attention relies on dynamic restructuring of cortical information flow to prioritize neuronal communication between those neuronal groups conveying information about behaviorally relevant information while reducing the influence from groups encoding irrelevant and distracting information. Electrophysiological evidence suggests that such selective neuronal communication is instantiated and sustained through selective neuronal synchronization of rhythmic gamma band activity within and between neuronal groups. Attentionally modulated synchronization patterns evolve rapidly, are evident even before sensory inputs arrive, follow closely subjective readiness to process information in time, can be sustained for prolonged time periods, and convey specific information about perceptually selected sensory features and motor plans. These functional implications of selective synchronization patterns are complemented by recent insights about the mechanistic origins of rhythmic synchronization at micro- and macro- scales of cortical neuronal processing, suggesting that selective attention is subserved by precise neuronal synchronization that is selective in space, time and frequency.

1 Introduction

Top–down attention is the key mechanism to restructure cortical information flow to prioritize processing of behaviorally relevant over irrelevant and distracting information (Gilbert and Sigman 2007). The behavioral consequences of attentional restructuring of information flow are manifold. Attended sensory inputs are

P. Fries (✉)
Donders Institute for Brain, Cognition and Behaviour, Radboud University Nijmegen, Nijmegen, The Netherlands
and
Ernst Strüngmann Institute (ESI) in Cooperation with Max Planck Society, Deutschordenstr. 46, 60528 Frankfurt, Germany
e-mail: pascal.fries@esi-frankfurt.de

S. Dehaene and Y. Christen (eds.), *Characterizing Consciousness: From Cognition to the Clinic?* Research and Perspectives in Neurosciences,
DOI 10.1007/978-3-642-18015-6_7, © Springer-Verlag Berlin Heidelberg 2011

processed more rapidly and accurately and with higher spatial resolution and sensitivity for fine changes, whereas non-attended information appears lower in contrast and is sometimes not perceived at all (Carrasco et al. 2004; Simons and Rensink 2005).

These functional consequences of attention require temporally dynamic and selective changes of neuronal interactions spanning multiple levels of neuronal information processing. Attentional selection (1) modulates interactions among single neurons within cortical microcircuits, (2) modulates the impact of selective local neuronal groups conveying relevant information within functionally specialized brain areas, and (3) controls long-range interactions among neuronal groups from distant brain areas (Maunsell and Treue 2006; Mitchell et al. 2007, 2009; Reynolds and Chelazzi 2004; Womelsdorf and Fries 2007, Cohen and Maunsell 2009).

For all these levels of neuronal interactions, converging evidence suggests that the selective modulation of interactions is critically built on selective synchronization. Neuronal synchronization is typically of an oscillatory nature, i.e., neurons fire and pause together in a common rhythm. When synchronization is rhythmic, it is often addressed as coherence, and we will use these terms interchangeably. This rhythmic synchronization can influence neuronal interactions in several ways: (1) spikes that are synchronized will have a larger impact on a target neuron than spikes that are not synchronized (Azouz and Gray 2003; Salinas and Sejnowski 2001); (2) local inhibition that is rhythmically synchronized leaves periods without inhibition, whereas non-synchronized inhibition will prevent local network activity continuously (Tiesinga et al. 2004); and (3) rhythmic synchronization of a local group of neurons will modulate the impact of input to that group, and therefore, the impact of rhythmic input will depend on the synchronization between input and target (Womelsdorf et al. 2007). These mechanisms are at work on all levels of attentional selection. At the level of microcircuits, inhibitory interneuron networks have been shown to impose rhythmic synchronization capable of effectively controlling the gain of the neuronal spiking output (Bartos et al. 2007; Tiesinga et al. 2008; Tiesinga and Buia 2009; Cardin et al. 2009). At the level of local neuronal groups, attention selectively synchronizes the responses of those neurons conveying information about the attended feature or location (Womelsdorf and Fries 2007). And the coherent output from these local neuronal groups has been shown to selectively synchronize over long-range connections with task-relevant neuronal groups in distant brain regions (Buschman and Miller 2007; Saalmann et al. 2007; Schoffelen et al. 2005; Sejnowski and Paulsen 2006; Pesaran et al. 2008; Gregoriou et al. 2009).

These empirical insights suggest that mechanisms underlying neuronal synchronization could be primary mechanisms behind selective attention. In particular, top–down attention may act by biasing rhythmic synchronization to establish and sustain a selective neuronal communication structure (Fries 2005). In this chapter, we begin by outlining this conceptual framework for selective attention through selective synchronization. We then survey basic insights from empirical and theoretical studies suggesting that rhythmic synchronization is particularly suited to control the selective routing of neuronal information flow, and we review how attention recruits these mechanisms across all levels of cortical processing.

2 Attentional Selection as a Dynamic Instantiation of a Selective Neuronal Communication Structure

During natural sensation, top–down control is dynamically established during ongoing processing. Experimentally, top–down signals are set by task instructions and by instructional cues defining relevant and irrelevant sensory features of the input stream during task performance. In typical paradigms of selective attention, the sensory input is kept identical across trials, with variations only in covert attention to different aspects of that input. In such tasks, neuronal responses are modulated with rapid temporal dynamics and high spatial selectivity throughout the cerebral cortex (Fig. 1a).

The temporal dynamics of attentional selection are illustrated by recent evidence of a rapid onset of selective neuronal response modulation in cortical areas as far apart as frontal cortex and primary visual cortices in the macaque brain (Khayat et al. 2006; Monosov et al. 2008; see also Gregoriou et al. 2009). In these studies, monkeys were instructed to detect a predefined target stimulus in visual displays to guide saccadic eye movement. In frontal and parietal cortex, attentional selection occurred within the first 120 ms following the sensory onset of target and distracter stimuli, allowing prediction of the spatial focus of attention (Gottlieb 2002; Monosov et al. 2008). As soon as about 30 ms later, top–down information changed neuronal responses at the earliest visual cortical processing stage in primary visual cortex (Khayat et al. 2006; Roelfsema et al. 2007), as was evident in a response enhancement for neurons with receptive fields overlapping the attentional target stimulus. These findings demonstrate that top–down control restructures cortical activity to sensory inputs across distant cortical sites on a rapid time scale. Attention amplifies almost instantaneously (i.e., with the sensory response latency) the influence of local groups of neurons conveying behavioral relevant information, and it attenuates the influence of neuronal groups coding for irrelevant inputs. This finding suggests that those distributed groups processing 'attended' inputs also interact effectively, establishing a selective neuronal communication structure on top of the existing infrastructure of anatomical connections (Fries 2005; Fig 1a). Interactions among neurons conveying information about attended locations or features are rendered effective, whereas anatomical connections between neuronal groups activated by distracting information are rendered ineffective.

Beyond the temporal dynamics of attentional selection, its spatial selectivity in restructuring cortical information flow is particularly evident across successive processing stages in visual cortex. Neurons at the highest visual processing stage in the inferotemporal cortex (IT) have receptive fields that span much of a visual field and respond selectively to complex objects composed of simpler visual features. Part of this selectivity arises from their broad and convergent anatomical input from neurons at earlier processing stages having smaller receptive fields and simpler tuning properties. During natural vision, the large receptive field of an IT neuron will typically contain multiple objects. However, when attention is directed to only one of those objects, the IT neuron's response is biased towards the response that

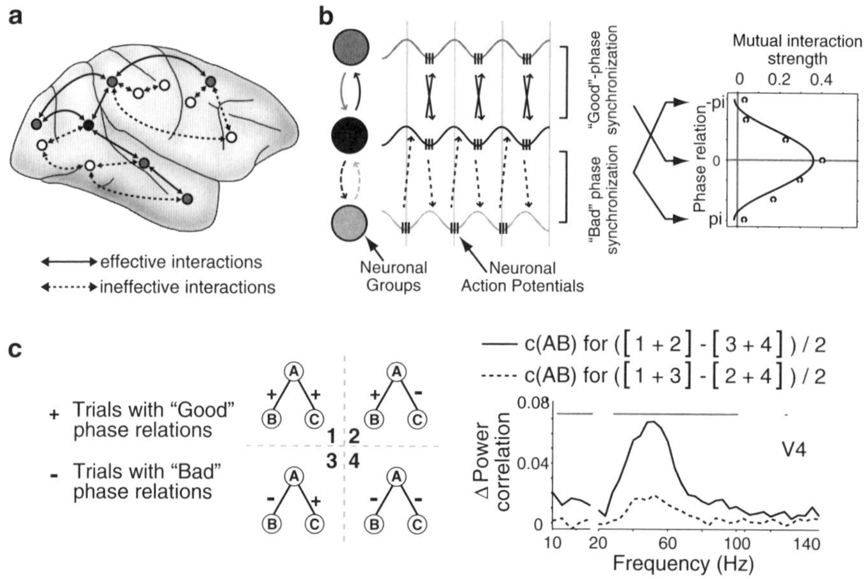

Fig. 1 Selective synchronization renders neuronal interactions among subsets of neuronal groups effective. (**a**) Anatomical connectivity (sketched as *lines*) provides a rich infrastructure for neuronal communication among neuronal groups (*circles*) throughout the cortex. With selective attention, only a small subset of these connections is rendered effective (*solid lines*). Interactions among groups conveying irrelevant information (*light gray circles*) for the task at hand are rendered less effective (*dashed lines*). (**b**) Illustration of the hypothesized role for selective synchronization in selective communication among three neuronal groups (*circles*). Rhythmic activity (LFP oscillations with spikes in troughs) provide briefly recurring time windows of maximum excitability (LFP troughs) that are either in-phase (*black and dark gray groups*) or in anti-phase (*black and light gray groups*). The plot on the right shows that mutual interactions (upper axis, correlation of the power of the LFP and the neuronal spiking response between neuronal groups) are high during periods of in-phase synchronization and lower otherwise. (**c**) The trial-by-trial interaction pattern between neuronal groups (A-to-B, and A-to-C) is predicted by the pattern of synchronization. If AB synchronizes at a good phase, their interaction is strongest, irrespective of whether A synchronizes with C at good or bad phase relations in the same trials. Thus, the spatial pattern of mutual interactions can be predicted by the phase of synchronization among rhythmically activated neuronal groups. Panels in (**b**) and (**c**) adapted from Womelsdorf et al. (2007)

would be obtained if only the attended object were presented (Chelazzi et al. 1993; Moran and Desimone 1985; Sejnowski and Paulsen 2006; Sheinberg and Logothetis 2001).

Such dynamic biasing of responses in IT cortex could be achieved by selective enhancement (suppression) of the impact of those afferent inputs from neurons in earlier visual areas coding for the attended (non-attended) input (Reynolds et al. 1999). However, the mechanisms underlying this up- and down-modulation of input gain for subsets of converging connections are only poorly understood; they likely entail a selective increase of temporally precise and coincident inputs from those neurons activated by an attended stimulus in earlier areas. This relevance of spike timing is suggested by fine-grained attentional modulation of precise neuronal synchronization within area V4 (Fries et al. 2001b; Bichot et al. 2005; Taylor et al. 2005; Womelsdorf et al. 2006; Chalk et al. 2010). Enhanced synchronization of the spiking output among those neuronal groups activated by attended sensory input (Fries et al. 2008) results in enhanced coincident arrival of their spikes at their postsynaptic target neurons in area IT. Temporally coincident input is highly effective in driving neuronal activity (Azouz and Gray 2003; Salinas and Sejnowski 2001; Tiesinga et al. 2008). It is therefore likely that selective synchronization within area V4 underlies attentional biasing within IT cortex and could thus underlie effective spatial routing of information flow within visual cortex.

Note that neuronal synchronization is, in principle, independent of firing rate, both in terms of metrics and physiology. The different metrics used for quantifying synchronization are typically normalized for firing rate. Physiologically, there are examples where enhanced firing rates are associated with strongly reduced synchronization, e.g., the stimulus-induced alpha-band desynchronization in the superficial layers of monkey V4 (Fries et al. 2008). Neuronal gamma-band synchronization typically emerges when neuronal groups are activated; therefore, it is in most cases associated with increased firing rates. However, firing rates and gamma-band synchronization can also be dissociated from each other; this can be found primarily when firing rate changes are not driven by changes in bottom-up input (e.g., stimulus changes) but rather by changes in top–down input (e.g., attention or stimulus selection; Fries et al. 2002; Womelsdorf et al. 2006).

Synchronization is a neuronal population phenomenon and it is often very difficult to assess it with recordings from isolated single units. Correspondingly, many studies of neuronal synchronization use recordings of multi-unit activity and/or of the local field potential (LFP). The LFP reflects the summed trans-membrane currents of neurons within a few 100 μm of tissue. Since synchronized currents sum up much more efficiently than unsynchronized currents, the LFP reflects primarily synchronized synaptic activity. Changes in LFP power typically correlate very well with changes in direct measures of neuronal synchronization.

Rhythmic synchronization within a neuronal group not only increases its impact on postsynaptic target neurons in a feedforward manner but also rhythmically modulates the group's ability to communicate, such that rhythmic synchronization between two neuronal groups likely subserves their interaction, because rhythmic inhibition within the two groups is coordinated and mutual inputs are optimally timed. We capture these implications in the framework of selective attention through selective synchronization (Fries 2005).

3 Selective Attention Through Selective Synchronization

Local neuronal groups frequently engage in periods of rhythmic synchronization. During activated states, rhythmic synchronization is typically evident in the gamma frequency band (30–90 Hz; Engel et al. 1990; Gray et al. 1989;

Hoogenboom et al. 2005). In vitro experiments and computational studies suggest that gamma-band synchronization emerges from the interplay of excitatory drive and rhythmic inhibition imposed by interneuron networks (Bartos et al. 2007; Börgers et al. 2005; Börgers and Kopell 2003; Buia and Tiesinga 2006). Interneurons impose synchronized inhibition onto the local network (Bartos et al. 2007; Hasenstaub et al. 2005; Vida et al. 2006). The brief time periods between inhibitions provide time windows for effective neuronal interactions with other neuronal groups, because they reflect enhanced postsynaptic sensitivity to input from other neuronal groups as well as maximal excitability for generating spiking output to other neuronal groups (Azouz 2005; Azouz and Gray 2003; Fries et al. 2007; Tiesinga et al. 2008). As a consequence, when two neuronal groups open their temporal windows for interaction at the same time, they will be more likely to mutually influence each other (Womelsdorf et al. 2007). The consequences for selective neuronal communication are illustrated in Fig. 1b: if the rhythmic synchronization within neuronal groups is precisely synchronized between the two groups, then they are maximally likely to interact. By the same token, if rhythmic activity within neuronal groups is uncorrelated between groups or synchronizes consistently out of phase, their interaction is curtailed (Fig. 1b).

This scenario requires that the pattern of synchronization between neuronal groups flexibly structures the pattern of interactions between neuronal groups (Fig. 1c). Consistent with this hypothesis, the interaction pattern of one neuronal group (A) with two other groups (B and C) can be predicted by their patterns of precise synchronization (Fig. 1c), as has recently been demonstrated for interactions of triplets of neuronal groups from within and between areas in awake cat and monkey visual cortex (Womelsdorf et al. 2007). This study measured the trial-by-trial changes in correlated amplitude fluctuation and changes in precise synchronization between pair AB and pair AC using the spontaneous variation of neuronal activity during constant visual stimulation. The strength of amplitude covariation, i.e., the covariation of power in the LFP and/or multiunit spiking responses, was considered the measure of mutual interaction strength. The results showed that the interaction strength of AB could be inferred from the phase of gamma-band synchronization between group A and group B being rather unaffected by the phase of synchronization of group A with group C (Fig. 1c). This finding was evident for triplets of neuronal groups spatially separated by as little as 650 μm, illustrating a high spatial resolution and specificity of the influence of precise phase synchronization between neuronal groups on the efficacy of neuronal interaction. Importantly, additional analysis supported a mechanistic role for the phase of synchronization between rhythmic activities to modulate the effective interaction strength (Womelsdorf et al. 2007). In particular, precise phase synchronization preceded higher amplitude covariations in time by a few milliseconds, arguing for a causal influence of precise phase synchronization to trigger neuronal interactions. Taken together, these results provide the most direct evidence available so far to suggest a critical mechanistic role of selective synchronization for neuronal interactions. They demonstrate that synchronization patterns can shape neuronal interactions with high specificity in time, space and frequency.

Importantly, these same characteristics of selective neuronal interactions are the key elements underlying selective attention. Attentional selection dynamically evolves at a rapid time scale and with high spatial resolution by enhancing (reducing) the effective connectivity among neuronal groups conveying task relevant (irrelevant) information. Such dynamic restructuring of neuronal interactions could be accomplished through mechanisms evoking selective synchronization patterns within interneuron networks. Selective changes of precise synchronization in local neuronal groups are capable of modulating in a self-emergent manner selective interaction patterns across neuronal groups (Börgers and Kopell 2008; Mishra et al. 2006; Tiesinga et al. 2008; Tiesinga and Buia 2009).

The outlined scheme of selective attention implemented as selective neuronal synchronization comprises explicit assumptions that selective attention affects interneuron networks and synchronization patterns during task performance. The following section surveys the available insights on interneuron networks and reviews the emerging signatures of attentional modulation of selective synchronization patterns in macaque cortex.

4 Synchronization in Interneuron Networks and Their Attentional Modulation

Interneurons comprise about a fifth of the neuron population but, despite their ubiquitous presence, their functional roles underlying cortical computations or cognitive processes are far from understood (Markram et al. 2004). However, a central role for the control of local cortical network activity has been suggested for the large class of interneurons of the basket cell type (Buzsaki 2006). These neurons target perisomatic regions of principal cells and are thereby capable of determining the impact of synaptic inputs arriving at sites distal to a cell's soma. Such perisomatic connectivity critically controls the input gain of principal cells across a large population of principal cells (Cardin et al. 2009; Tiesinga and Sejnowski 2009; Buzsaki et al. 2007; Cobb et al. 1995; Markram et al. 1998; Rudolph et al. 2007; Tiesinga et al. 2004). As described above, the inhibitory synaptic influence is inherently rhythmic at high frequencies, carrying stronger gamma band power than pyramidal cells (Cardin et al. 2009; Bartos et al. 2007; Hasenstaub et al. 2005).

The prominent role of these high frequency inputs in shaping the spiking output of principal cells has recently been demonstrated directly in cat and rodent visual cortex. It was shown that the spiking of principal cells is indeed preceded by brief periods of reduced inhibition (Rudolph et al. 2007; see also Fig. 8 of Hasenstaub et al. 2005 and Fig. 4 of Cardin et al. 2009). Taken together, these findings suggest that interneurons are the source of rhythmic inhibition onto a local group of neurons synchronizing the discharge of pyramidal cells to the time windows between inhibitions.

In the context of selective attention, interneuron networks could be activated by various possible sources. They may be activated by transient, and spatially specific, neuromodulatory inputs (Lin et al. 2006; Rodriguez et al. 2004). Alternatively, selective attention could target local interneuron networks directly via top–down inputs from neurons in upstream areas (Buia and Tiesinga 2008; Tiesinga and Buia 2009; Mishra et al. 2006; Tiesinga et al. 2008). In these models, selective synchronization emerges either by depolarizing selective subsets of interneurons (Buia and Tiesinga 2008; Tiesinga and Sejnowski 2004) or by biasing the phase of rhythmic activity in a more global inhibitory interneuron pool (Mishra et al. 2006). In either case, rhythmic inhibition controls the spiking responses of groups of excitatory neurons, enhancing the impact of those neurons spiking synchronously within the periods of disinhibition while actively reducing the impact of neurons spiking asynchronously to this rhythm. This suppressive influence on excitatory neurons, which are activated by distracting feedforward inputs, reflects the critical ingredient for the concept of selective attention through selective synchronization: attention not only enhances synchronization of already more coherent activity representing attended stimuli but also actively suppresses the synchronization and impact of groups of neurons receiving strong, albeit distracting inputs, because they arrive at non-optimal phase relations to the non-inhibited periods in the target group. The computational feasibility of both facilitatory and suppressive aspects, and the critical role of the timing of inhibitory circuits, have recently received direct support (Börgers and Kopell 2008; Tiesinga and Buia 2009).

Despite the prominent computational role of interneuron activity for selective communication, there are only sparse insights into its implications in selective information processing during cognitive task performance. The basic prediction from the above models is that interneurons are attentionally modulated. Consistent with this presupposition, a recent study by Mitchell et al. (2007) reports a clear attentional modulation of putative interneurons in visual area V4 during a selective attention task requiring monkeys to track moving grating stimuli. Putative interneurons showed similar relative increases in firing rate and greater increases in reliability compared to putative pyramidal neurons. However, tests of more refined predictions about the relative modulation of synchronization and the phase relation of spiking responses of inhibitory and excitatory neuron types still need to be conducted (Buia and Tiesinga 2008).

5 Selective Modulation of Synchronization During Attentional Processing

Direct evidence for the functional significance of selective synchronization within local neuronal groups for attentional selection has been obtained from recordings in macaque visual cortical area V4 (Fries et al. 2001a, b; Taylor et al. 2005; Womelsdorf et al. 2006; Chalk et al. 2010; Gregoriou et al. 2009). One consistent result across studies in V4 is that spatial attention enhances gamma-band synchronization within those neuronal groups with receptive fields overlapping the attended location. The enhanced rhythmic synchronization is strongly evident within the LFP signal and in more precise synchronization of neuronal spiking responses to the LFP. Importantly, the synchronization among the spiking output from neurons coding for the attended location is also enhanced compared to the spiking output of neurons activated by a non-attended distractor stimulus (Fig. 2; Fries et al. 2008). These attentional effects on spike-to-spike synchronization imply that the

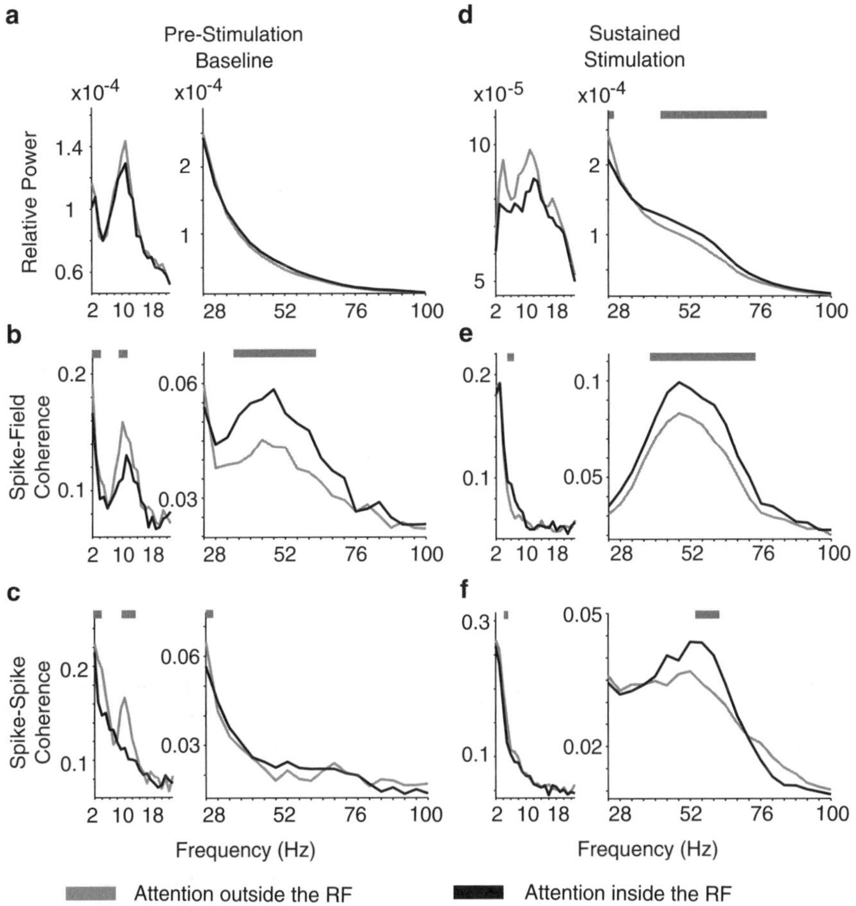

Fig. 2 The pattern of attentional modulation of synchronization in macaque visual area V4 before and during sensory stimulation. (**a–c**) Attentional modulation of relative LFP power (**a**), spike-to-LFP coherence (**b**) and spike-to-spike coherence (**c**) across low and high frequencies during the baseline period of a spatial attention task. Monkeys either attended (*dark lines*) or ignored (*gray lines*) the receptive field (RF) location of the recorded neuronal groups in blocks of trials. (**d–f**) Attentional modulation of the neuronal response during stimulation with an attended/ignored moving grating. Same format as in (**a–c**). Horizontal gray bars denote frequencies with significant attentional effects. Adapted from Fries et al. (2008)

postsynaptic targets receive more coherent input from those neuronal groups conveying behaviorally relevant information.

5.1 Functional Implications of Selective Gamma-Band Synchronization

In addition to the described attentional effect, recent studies demonstrated that the precision of local synchronization in visual area V4 is closely related to task performance, including behavioral accuracy and the reaction time to detect behaviorally relevant stimulus changes (Taylor et al. 2005; Womelsdorf et al. 2006). The relation to behavioral accuracy was derived from an error analysis of the pattern of synchronization in area V4 (Taylor et al. 2005). In this study, the spatial focus of attention could be inferred from the pattern of synchronization measured through epidural electrodes. Gamma-band synchronization was not only stronger for correct trials than for miss trials but also the degree of synchronization predicted whether the monkey was paying attention to the distracter. Thus, this study demonstrated that gamma-band synchronization reflects the actual allocation of attention rather than merely the attentional cueing itself. The link to reaction time was made in a recent study, which demonstrated that the precision of stimulus-induced gamma-band synchronization predicts how rapidly a stimulus change can be reported behaviorally. When monkeys were spatially cued to select one of two stimuli to detect a color change of the attended stimulus, the speed of change detection could be partly predicted by the strength of gamma-band synchronization shortly before the stimulus change actually occurred (Womelsdorf et al. 2006). Importantly, the reaction times to the stimulus change could not be predicted at times before the stimulus change by overall firing rates or by synchronization outside of the gamma band. Notably, the correlation of gamma-band synchronization with the speed of change detection showed high spatial selectivity: neurons activated by an unattended stimulus engaged in lower synchronization when the monkeys were particularly fast in responding to the stimulus change at locations outside their receptive field. This finding rules out a possible influence of globally increased synchronization during states of enhanced alertness and arousal (Herculano-Houzel et al. 1999; Munk et al. 1996; Rodriguez et al. 2004), and it argues for a fine-grained influence of synchronization to modulate the effective transmission of information about the stimulus change to postsynaptic target areas concerned with the planning and execution of responses.

These behavioral correlates of gamma-band synchronization during selective attention tasks are complemented by a variety of correlational results linking enhanced gamma-band synchronization to efficient task performance in various attention-demanding paradigms. For example, in memory-related structures, the strength of gamma band synchronization has been linked to the successful encoding and retrieval of information (Montgomery and Buzsaki 2007; Sederberg et al. 2003, 2006a, b; Jutras et al. 2009).

5.2 *Selective Gamma-Band Coherence Beyond Visual Cortex*

These results of selective gamma-band synchronization with selective spatial attention are supported by a growing number of converging findings from human EEG and MEG studies (Doesburg et al. 2008; Fan et al. 2007; Wyart and Tallon-Baudry 2008; Siegel et al. 2008). Importantly, attention modulates gamma-band synchronization beyond sensory visual cortex. It has been reported for auditory cortex (Kaiser et al. 2006; Tiitinen et al. 1993) and more recently in somatosensory cortex (Bauer et al. 2006; Hauck et al. 2007). Spatial attention for tactile discrimination at either the right or left index finger in humans enhanced stimulus-induced gamma-band synchronization in primary somatosensory cortex when measured with MEG. Similar topographies and dynamics of gamma-band synchronization correlate with the actual perception of somatosensory induced pain (Gross et al. 2007). Importantly, enhanced oscillatory dynamics in the gamma band during tactile perception is not restricted to the somatosensory cortex (Ohara et al. 2006). In recent intracranial recordings in humans, synchronization was modulated across somatosensory cortex, medial prefrontal and insular regions when subjects had to direct attention to painful tactile stimulation (Ohara et al. 2006).

5.3 *Spatially Specific Synchronization Patterns During Preparatory Attentional States*

The described gamma-band modulation of rhythmic activity is most prominent during activated states. However, attentional top–down control biases neuronal responses in sensory cortices even before sensory inputs impinge on the neuronal network (Fries et al. 2001b, 2008; Luck et al. 1997; Siegel et al. 2008; Schroeder et al. 2010). In many attention studies, the instructional cue period is followed by a temporal delay void of sensory stimulation. During these preparatory periods, top–down signals set the stage for efficient processing of expected stimulus information, rendering local neuronal groups ready to enhance the representation of attended sensory inputs. Intriguingly, the described preparatory bias is evident in selective synchronization patterns in the gamma band.

In macaque visual cortical area V4, neurons gamma synchronize their spiking responses to the LFP more precisely when monkeys expected a target stimulus at the receptive field location of the respective neuronal group (Fig. 2b). This modulation was evident even though rhythmic activity proceeded at far lower levels in the absence of sensory stimulation compared to synchronization strength during high contrast sensory drive. Lower overall strength, and correspondingly lower signal-to-noise ratio, may account for the lack of significant gamma-band modulation of LFP power or spike-to-spike synchronization during the pre-stimulus period when compared to attentional modulation during stimulation (Fig. 2).

During preparatory periods, and thus in the absence of strong excitatory drive to the local network, rhythmic activity is dominated by frequencies lower than the gamma band. In the described study from macaque V4, pre-stimulus periods were characterized by alpha-band peaks of local rhythmic synchronization when monkeys attended away from the receptive field of the neuronal group. Figure 2b, c demonstrates reduced locking of neuronal spiking in the alpha band to the LFP and to spiking output of nearby neurons. This finding agrees with various studies reporting reduced alpha-band activity during attentional processing (Bauer et al. 2006; Pesaran et al. 2002; Rihs et al. 2007; Worden et al. 2000; Wyart and Tallon-Baudry 2008; Siegel et al. 2008). Interestingly, human EEG studies extend this finding by showing that the degree of alpha frequency desynchronization during prestimulus intervals of visuo-spatial attention tasks indicates how fast a forthcoming target stimulus is processed (Jin et al. 2006; Sauseng et al. 2006; Thut et al. 2006). For example, reaction times to a peripherally cued target stimulus are partially predicted by the lateralization of alpha activity in a 1-s period before target appearance (Thut et al. 2006). While this predictive effect was based predominantly on reduced alpha-band responses over the hemisphere processing the attended position, recent studies suggest that alpha-band oscillations are selectively enhanced within local neuronal groups processing distracting information, i.e., at unattended locations (Kelly et al. 2006; Rihs et al. 2007; Yamagishi et al. 2003). These findings suggest that rhythmic alpha-band synchronization may play an active role in preventing the signaling of stimulus information. According to this hypothesis, attention up-regulates alpha-band activity of neuronal groups expected to process distracting stimulus information rather than to down-regulate local alpha-band synchronization for neuronal groups processing attended stimulus features and locations.

5.4 Synchronization Patterns Reflecting Temporal Expectancies of Target Processing

The previous paragraph surveyed evidence for an influence of spatially specific expectancy of target and distracter stimuli on synchronization patterns in visual cortex. In addition to spatially selective expectancy, the expectation of the occurrence of behaviorally relevant target events is known to influence neuronal synchronization patterns and firing rates in parietal and frontal cortex (Ghose and Maunsell 2002; Janssen and Shadlen 2005; Riehle 2005; Schoffelen et al. 2005; Pesaran et al. 2008; Gregoriou et al. 2009; Schroeder and Lakatos 2009). Attentional modulation of neuronal firing rates in extrastriate visual area MT (middle temporal) is strongest around the time point at which the subjective anticipation for a target change, given that it had not occurred before in the trial (i.e., the hazard rate), is maximal (Ghose and Maunsell 2002). In premotor and motor cortex, the hazard rate is smoothly reflected in the strength of synchronization (Riehle 2005;

Schoffelen et al. 2005). Importantly, enhanced readiness to respond to attended sensory changes is thereby functionally closely linked to long-range synchronization of motor cortex with spinal motor units, suggesting a direct mechanistic influence of synchronization on the speed to respond to behaviorally relevant sensory events (Schoffelen et al. 2005).

An influence of temporal expectancy on synchronization in early sensory cortices has recently been demonstrated in recordings in primary visual cortex of macaques (Lakatos et al. 2008). In this study, monkeys were cued to detect deviant sensory stimuli in either an auditory or visual input stream to receive reward. Auditory and visual stimuli alternated, and both stimulus streams followed a noisy 1.55 Hz rhythm. This low frequency rhythm of sensory inputs entrained neuronal responses in early visual cortex, such that responses to individual stimuli in the visual stream added to the entrained response. Attention to the visual stream amplified the entrainment (Fig. 3a), but the most prominent attentional effect was evident in the phase of the 1.55 Hz entrainment in the superficial layers of visual cortex. This entrainment was always determined by the stimulus stream that was attended, i.e., it switched by half a cycle when attention switched from the visual to

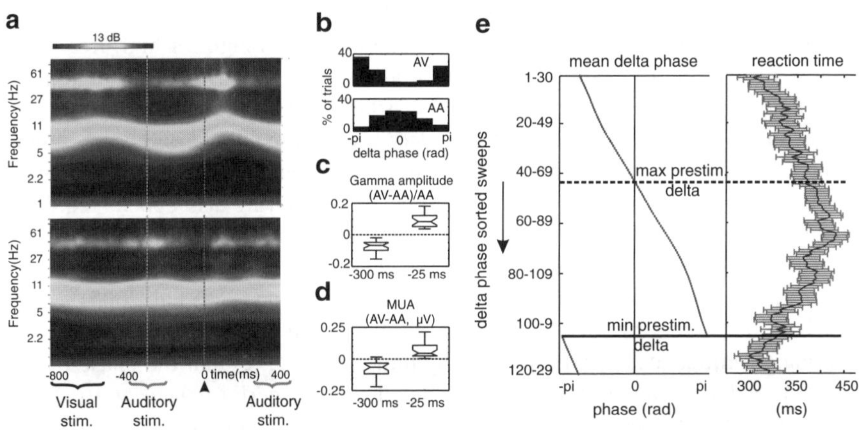

Fig. 3 (**a**) Entrainment of synchronization from delta- to gamma-band frequencies in supragranular layers of the primary visual cortex during an auditory-visual change detection task. (**b**) Time-frequency spectrograms during attention to the visual (*upper panel*) and auditory (*bottom panel*) input stream, aligned to the onset time of the visual stimulus. The task cued monkeys to detect infrequent deviant stimuli in either the auditory (*white noise tones*) or visual (*red light flashes*) input stream. Visual (auditory) stimuli were onset at a regular interval of 650 $\pm$ 150 ms, as indicated below the time axis (mean stimulus rate = 1.55 Hz). (**c**) Entrainment of delta frequency phase in visual cortex measured as the delta (1.55 Hz) phase at the time of visual stimulus onset, when attention was directed to the visual (*upper panel*) and auditory (*lower panel*) modality across recording sessions. (**d**, **e**) Modulation of gamma-band amplitude of the LFP (**d**) and multiunit activity or MUA (**e**) before and at visual stimulus onset. Positive values indicate enhancement with visual versus auditory attention. (**f**) Reaction times (x-axis) to the visual target stimulus sorted into groups of trials according to the pre-stimulus delta phase (at 0 ms to stimulus onset; y-axis). *Solid/dashed horizontal lines* indicate the group of trials corresponding to maximum/minimum delta amplitudes. Adapted from Lakatos et al. (2008)

the auditory stream (Fig. 3b), which had a phase opposite to the visual stream. Importantly, low frequency fluctuations in the LFP likely reflect fluctuations in neuronal excitability. With attention to the visual (auditory) input stream, the phase corresponding to maximal (minimal) neuronal excitability occurred around the average time when the target information was most likely to reach visual cortex. Consistent with a functional role of the entrained delta phase, the authors reported the strongest attentional enhancement of gamma-band synchronization in the LFP and spiking activity around this time (Fig. 3c, d) and showed that the detection of deviant visual stimuli was fastest (slowest) when the delta phase at stimulus onset corresponded to maximal (minimal) neuronal excitability (Fig. 3e).

The described results suggest that top–down information selectively modulates excitability in early sensory cortices through changes in the phase of rhythmic entrainment in these areas (Lakatos et al. 2009). The exact frequency band underlying excitability modulations may extend from the low delta band, directly imposed by the stimulus structure in the described study, to the theta band around 4–8 Hz. This suggestion may be derived from the time-frequency evolution of LFP power in the theta band and its attentional modulation, as shown in Fig. 3a. Intriguingly, similar to the effect of delta phase on the gamma-band response demonstrated directly in the discussed study (Fig. 3b), previous studies have linked the phase of rhythmic activity in the theta band to the strength of high-frequency gamma-band synchronization in rodent hippocampus and over large regions in the human cortex (Canolty et al. 2006; Csicsvari et al. 2003). An additional hint suggesting a functional relevance of low-frequency phase fluctuations can be found in a recent study, demonstrating that spiking responses in rodent prefrontal cortex phase lock to theta-band activity in the hippocampus during task epochs requiring spatial decisions in a working memory context (Jones and Wilson 2005). In macaque visual cortex, the phase of theta-band synchronization has been directly linked to selective maintenance of task-related information (Lee et al. 2005). Taken together, the emerging evidence demonstrates (1) that top–down, task-related information modulates low-frequency rhythmic activity, (2) that the phase of this rhythmic activity can be functionally related to task performance, and (3) that the phase of low-frequency activity shapes the strength of gamma-band synchronization in response to sensory inputs. As such, the pattern of selective synchronization in the gamma band described in the previous paragraphs could be tightly linked to underlying, selective low-frequency activity modulations. Whether both are coupled in an obligatory way, or whether the co-modulation may be triggered by specific task demands, will be an interesting subject for future research (Schroeder et al. 2010).

5.5 *Feature-Selective Modulation of Rhythmic Synchronization*

The preceding sections discussed evidence for selective neuronal synchronization patterns evolving with space-based attentional selection of sensory inputs. However, in addition to spatial selection, attention frequently proceeds only on top–down information about the behaviorally relevant sensory feature and independent of the exact spatial location at which input impinges on sensory cortices. Such feature-based attention is known to modulate the responses of neurons tuned to the attended feature, such as a particular motion direction or the color of a visual stimulus (Maunsell and Treue 2006).

Importantly, a recent study demonstrated that attention to a particular feature selectively synchronizes the responses of neurons tuned to the attended stimulus feature (Bichot et al. 2005). In this study, spiking responses and LFPs were recorded in macaque visual area V4 while monkeys searched in multi-stimulus displays for a target stimulus defined either by color, shape or both. When monkeys searched for a red stimulus, for example, by shifting their gaze across stimuli on the display, the non-foveal receptive fields of the recorded neurons could either encompass non-target stimuli (e.g., of blue color) or the (red) target stimulus prior to the time when the monkey detected the target. The authors found that neurons synchronized to the LFP more strongly in response to their preferred stimulus feature when it was the attended search target feature rather than a distracter feature.

Thus, attention enhanced synchronization of the responses of those neurons sharing a preference for the attended target feature, irrespective of the spatial location of attention (Bichot et al. 2005). This feature-based modulation was also evident during a conjunction search task involving targets defined by two features: (1) when monkeys searched for a target stimulus with a particular orientation and color (e.g., a red horizontal bar), neurons with preference to one of these features enhanced their neuronal synchronization (Bichot et al. 2005), and (2) this enhancement was observed not only in response to the color-shape defined conjunction target but also in response to distracters sharing one feature with the target (e.g., red color). This latter finding corresponds well with the behavioral consequences of increased difficulty and search time needed for conjunction-defined targets.

This study shows that feature salience is indexed not only by changes in firing rates (Martinez-Trujillo and Treue 2004; Treue and Martinez Trujillo 1999; Wannig et al. 2007) but also by selectively synchronizing neuronal responses, depending on the similarity between neuronal feature preferences and the attended stimulus feature. The mechanisms behind this selective influence of featural top–down information could be based on a similar spatial weighting of interneuron network activity as implicated for spatial selection. Neuronal tuning to many basic sensory features is organized in regularly arranged local maps. Correspondingly, the tuning of groups of neurons measured with the LFP is locally highly selective. Importantly, neuronal stimulus preference is systematically related to the strength of neuronal synchronization in the gamma-frequency band, which has been demonstrated for stimulus orientation and spatial frequency (Frien et al. 2000; Gray et al. 1990; Kayser and König 2004; Kreiter and Singer 1996; Siegel and König 2003), the speed and direction of visual motion (Liu and Newsome 2006), and the spatial motor intentions and movement directions (Scherberger and Andersen 2007; Scherberger et al. 2005). These findings show that rhythmic synchronization conveys feature-selective information. Feature-based attention appears to recruit this property with high spatial resolution by modulating which neurons synchronize to the local rhythmic activity.

Taken together, the previous subsections surveyed the accumulating evidence, demonstrating selective neuronal synchronization patterns that evolve with selective spatial and feature-based attention within sensory cortices. Only few studies have extended these insights to investigate how selective attention modulates selective neuronal interaction patterns between different visual areas and between visual and higher-order cortical areas during task performance. Recent evidence shows that such dynamic inter-areal interaction patterns are evident in long-range synchronization patterns between cortical areas.

5.6 Selective Inter-Areal Synchronization During Attentional Processing

In the preceding sections, selective synchronization patterns evolved for local neuronal groups in sensory cortices, supporting a functional role for gamma-band synchronization in the selective restructuring of neuronal communication during attentional processing (Fig. 1). However, attentional processing relies on effective interactions between local subsets of neuronal groups from distant cortical regions. So far, only few studies have investigated these inter-areal interaction patterns during task epochs with selective attention (Engel et al. 2001; Varela et al. 2001; Womelsdorf and Fries 2007). The emerging evidence from these studies points towards a critical role for rhythmic long-range synchronization in the gamma band (Gregoriou et al. 2009; see Fig. 4) and at lower frequencies, most prominently at beta frequencies ranging from 15 to 30 Hz.

Early studies in awake cats demonstrated transiently enhanced beta-frequency synchronization among visual cortical and premotor regions and between visual cortex and thalamus during non-selective states of expectancy of a behaviorally relevant stimulus (in, e.g., 'Go/No-Go tasks'; Roelfsema et al. 1997; von Stein et al. 2000; Wrobel et al. 2007). Recent studies in the macaque monkey have extended these findings by showing that fronto-parietal and intra-parietal interactions between areas are accompanied by synchronization at beta frequencies (15–35 Hz) during task epochs, requiring searching for and selecting behaviorally relevant visual stimuli (Buschman and Miller 2007; Saalmann et al. 2007; Pesaran et al. 2008). Figure 5 illustrates findings from a visual search task requiring monkeys to detect a search target that is either salient and pops out among distracting stimuli ("bottom–up search") or that is non-salient by sharing features with distracting stimuli (Buschman and Miller 2007). In contrast to bottom–up salient targets, the non-salient target stimuli were detected more slowly, indicating that they require attentive search through the stimuli in the display before they are successfully detected ('top–down search'). Paralleling the difference in behavioral

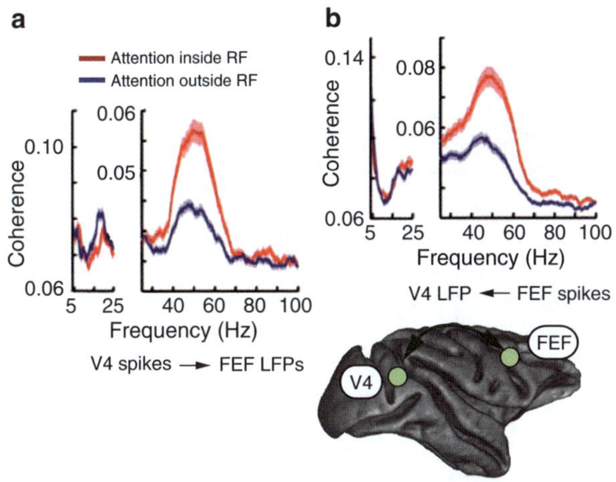

Fig. 4 Selective modulation of long-range synchronization between the frontal eye field (FEF) and visual area V4 during states of selective visual attention (Gregoriou et al. 2009). (**a**) Coherence (y-axis) of spike trains in V4 to the local field potential (LFP) in FEF, across frequency (x-axis), when attention was directed to a stimulus inside (*red*) and outside (*blue*) the receptive field (V4) and response field (FEF). Recording areas are indicated in the macaque brain outline below the panel. (**b**) Same as in **a**, but depicting the coherence between spike trains in FEF and the LFP in V4. Adapted from Gregoriou et al. (2009)

demands, the authors found a selective synchronization pattern among the LFPs in frontal and parietal cortex. While attentive 'top–down search' specifically enhanced rhythmic synchronization at 20–35 Hz compared to the "bottom–up" search, the stimulus-driven 'bottom–up' search resulted in stronger inter-areal synchronization in the gamma-frequency band (Fig. 5b). This pattern of results is most likely due to relative differences in task demands in both search modes and was unaffected by differences in reaction times. Therefore, these findings suggest that inter-areal communication during attentional top–down control is conveyed particularly through rhythmic synchronization in a high beta band, either in addition to or separate from the frequency of rhythmic interactions underlying bottom–up feedforward signaling (Engel and Fries 2010).

Consistent with a functional role for top–down mediated, long-range neuronal communication, various experimental paradigms demanding attentive processing have shown long-range synchronization in a broad beta band, although mostly at frequencies below 25 Hz. The following provide a few examples of beta-band modulation in recent studies using very different task paradigms. Freely choosing a sequence of target stimuli for arm movements induces 15 Hz coherence among neurons in premotor cortex and the parietal reach region when compared to instructed searches (Pesaran et al. 2008). Notably, during periods of enhanced inter-areal coherence, spiking activity in premotor cortex was more predictive of the direction of forthcoming arm movements compared to periods of lower coherence (Pesaran et al. 2008). Variations in reaction times and readiness to

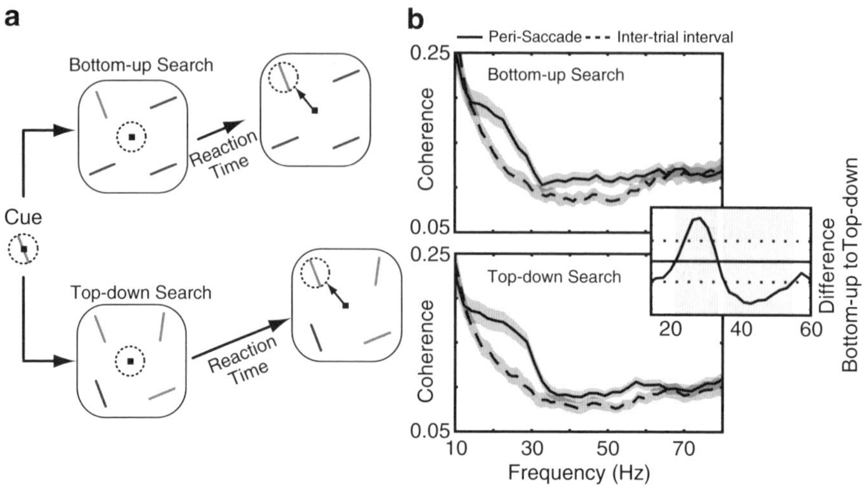

Fig. 5 Selective modulation of long-range synchronization between frontal and parietal cortex during visual search. (**a**) Sketch of two visual search tasks used by Buschman and Miller (2007). A cue instructed monkeys about the orientation and color of a bar that was the later search target in a multi-stimulus display during a bottom-up search task (both target color and orientation were unique; *upper panels*) and during a top–down search task (target shared color or orientation with distracting stimuli; *bottom panels*). Monkeys covertly attended the multi-stimulus array and made a saccade to the target stimulus position as soon as they found it. (**b**) The authors measured the coherence of the LFP activity of neuronal groups in the frontal eye field and dorso-lateral prefrontal cortex and parietal area LIP. The line plots on the right show the coherence (y-axis) for different frequency bands (x-axis) in the bottom–up and top–down tasks, along with the coherence difference across tasks (*solid line* in inset). The results show that attentional demand modulated long-range fronto-parietal coherence at different frequency bands. Adapted from Buschman and Miller (2007)

respond to a sensory change event induced corresponding fine-grained variations of motor-spinal coherence in the beta band (Schoffelen et al. 2005). Somatosensory and motor cortex synchronize in the beta band during sensorimotor integration (Brovelli et al. 2004). Selective working memory maintenance in a delayed match-to-sample task results in stronger coherence in the beta band between higher visual areas in humans (Tallon-Baudry et al. 2001) and locally predicts performance in a similar task in the monkey (Tallon-Baudry et al. 2004). Failure to detect a target stimulus in a rapid stream of stimuli in the attentional blink paradigm is associated with reduced fronto-parietal and fronto-temporal beta-band synchronization (Gross et al. 2004). And, as a last example of a potential functional role for beta-band activity, the perception of coherent objects from fragmented visual scenes goes along with transiently enhanced beta-band synchronization of the LFP among prefrontal, hippocampal and lateral occipital sites (Sehatpour et al. 2008).

Taken together, these diverse findings agree to suggest that inter-areal synchronization critically subserves neuronal interactions during attentive processing. In the surveyed studies, synchronization in a broadly defined beta band occurred selectively during task epochs requiring effective neuronal integration of information across distributed cortical areas. However, further studies need to elucidate the properties of particular frequency bands and their characteristic recruitment during specific tasks (Kopell et al. 2000).

6 Concluding Remarks

Selective attention describes a central, top–down process that restructures neuronal activity patterns to establish a selective representation of behavioral relevance. The surveyed evidence suggests that attention achieves this functional role by selectively synchronizing those neuronal groups conveying task-relevant information. Attentionally modulated synchronization patterns evolve rapidly, are evident even before sensory inputs arrive, follow closely subjective readiness to process information in time, can be sustained for prolonged time periods, and carry specific information about top–down selected sensory features and motor aspects.

In addition to these functional characteristics, insights into the physiological origins of synchronization have begun to shed light on the mechanistic underpinning of selective neuronal interaction patterns at all spatial scales of cortical processing. At the level of single neurons and local microcircuits, studies are deciphering the role of inhibitory interneuron networks, how precise timing information is conveyed and sustained even at high oscillation frequencies, and how rhythmic synchronization among interneurons is actively made robust against external influences (Bartos et al. 2007; Vida et al. 2006). These insights are integrated at the network level in models demonstrating how selective synchronization patterns evolve in a self-organized way (Börgers and Kopell 2008; Tiesinga et al. 2008). Acknowledging those basic physiological processes underlying the dynamic generation of selective synchronization seems to be pivotal to further elucidating the mechanistic working principles of selective attention in the brain.

Acknowledgments This work was supported by the European Science Foundation's European Young Investigator Award program (P.F.), by the Netherlands Organization for Scientific Research (P.F. and T.W.), and by the Canadian Institutes of Health Research (T.W.).

References

Azouz R (2005) Dynamic spatiotemporal synaptic integration in cortical neurons: neuronal gain, revisited. J Neurophysiol 94:2785–2796

Azouz R, Gray CM (2003) Adaptive coincidence detection and dynamic gain control in visual cortical neurons in vivo. Neuron 37:513–523

Bartos M, Vida I, Jonas P (2007) Synaptic mechanisms of synchronized gamma oscillations in inhibitory interneuron networks. Nat Rev Neurosci 8:45–56

Bauer M, Oostenveld R, Peeters M, Fries P (2006) Tactile spatial attention enhances gamma-band activity in somatosensory cortex and reduces low-frequency activity in parieto-occipital areas. J Neurosci 26:490–501

Bichot NP, Rossi AF, Desimone R (2005) Parallel and serial neural mechanisms for visual search in macaque area V4. Science 308:529–534

Börgers C, Kopell NJ (2003) Synchronization in networks of excitatory and inhibitory neurons with sparse, random connectivity. Neural Comput 15:509–538

Börgers C, Kopell NJ (2008) Gamma oscillations and stimulus selection. Neural Comput 20:383–414

Börgers C, Epstein S, Kopell NJ (2005) Background gamma rhythmicity and attention in cortical local circuits: a computational study. Proc Natl Acad Sci USA 102:7002–7007

Brovelli A, Ding M, Ledberg A, Chen Y, Nakamura R, Bressler SL (2004) Beta oscillations in a large-scale sensorimotor cortical network: directional influences revealed by Granger causality. Proc Natl Acad Sci USA 101:9849–9854

Buia C, Tiesinga P (2006) Attentional modulation of firing rate and synchrony in a model cortical network. J Comput Neurosci 20:247–264

Buia CI, Tiesinga PH (2008) The role of interneuron diversity in the cortical microcircuit for attention. J Neurophysiol 99:2158–2182

Buschman TJ, Miller EK (2007) Top-down versus bottom-up control of attention in the prefrontal and posterior parietal cortices. Science 315:1860–1862

Buzsaki G (2006) Rhythms of the brain. Oxford University Press Inc., Oxford

Buzsaki G, Kaila K, Raichle M (2007) Inhibition and brain work. Neuron 56:771–783

Canolty RT, Edwards E, Dalal SS, Soltani M, Nagarajan SS, Kirsch HE, Berger MS, Barbaro NM, Knight RT (2006) High gamma power is phase-locked to theta oscillations in human neocortex. Science 313:1626–1628

Cardin JA, Carlen M, Meletis K, Knoblich U, Zhang F, Deisseroth K, Tsai LH, Moore CI (2009) Driving fast-spiking cells induces gamma rhythm and controls sensory responses. Nature 459:663–667

Carrasco M, Ling S, Read S (2004) Attention alters appearance. Nat Neurosci 7:308–313

Chalk M, Herrero JL, Gieselmann MA, Delicato LS, Gotthardt S, Thiele A (2010) Attention reduces stimulus-driven gamma frequency oscillations and spike field coherence in V1. Neuron 66:114–125

Chelazzi L, Miller EK, Duncan J, Desimone R (1993) A neural basis for visual search in inferior temporal cortex. Nature 363:345–347

Cobb SR, Buhl EH, Halasy K, Paulsen O, Somogyi P (1995) Synchronization of neuronal activity in hippocampus by individual GABAergic interneurons. Nature 378:75–78

Cohen MR, Maunsell JH (2009) Attention improves performance primarily by reducing interneuronal correlations. Nat Neurosci 12:1594–1600

Csicsvari J, Jamieson B, Wise KD, Buzsaki G (2003) Mechanisms of gamma oscillations in the hippocampus of the behaving rat. Neuron 37:311–322

Doesburg SM, Roggeveen AB, Kitajo K, Ward LM (2008) Large-scale gamma-band phase synchronization and selective attention. Cereb Cortex 18:386–396

Engel AK, Fries P (2010) Beta-band oscillations-signalling the status quo? Curr Opin Neurobiol 20:156–165

Engel AK, Konig P, Gray CM, Singer W (1990) Stimulus-dependent neuronal oscillations in cat visual cortex: inter-columnar interaction as determined by cross-correlation analysis. Eur J Neurosci 2:588–606

Engel AK, Fries P, Singer W (2001) Dynamic predictions: oscillations and synchrony in top-down processing. Nat Rev Neurosci 2:704–716

Fan J, Byrne J, Worden MS, Guise KG, McCandlissa BD, Fossella J, Posner MI (2007) The relation of brain oscillations to attentional networks. J Neurosci 27:6197–6206

Frien A, Eckhorn R, Bauer R, Woelbern T, Gabriel A (2000) Fast oscillations display sharper orientation tuning than slower components of the same recordings in striate cortex of the awake monkey. Eur J Neurosci 12:1453–1465

Fries P (2005) A mechanism for cognitive dynamics: neuronal communication through neuronal coherence. Trends Cogn Sci 9:474–480
Fries P, Neuenschwander S, Engel AK, Goebel R, Singer W (2001a) Rapid feature selective neuronal synchronization through correlated latency shifting. Nat Neurosci 4:194–200
Fries P, Reynolds JH, Rorie AE, Desimone R (2001b) Modulation of oscillatory neuronal synchronization by selective visual attention. Science 291:1560–1563
Fries P, Schröder JH, Roelfsema PR, Singer W, Engel AK (2002) Oscillatory neuronal synchronization in primary visual cortex as a correlate of stimulus selection. J Neurosci 22:3739–3754
Fries P, Nikolic D, Singer W (2007) The gamma cycle. Trends Neurosci 30:309–316
Fries P, Womelsdorf T, Oostenveld R, Desimone R (2008) The effects of visual stimulation and selective visual attention on rhythmic neuronal synchronization in macaque area V4. J Neurosci 28:4823–4835
Ghose GM, Maunsell JH (2002) Attentional modulation in visual cortex depends on task timing. Nature 419:616–620
Gilbert CD, Sigman M (2007) Brain states: top-down influences in sensory processing. Neuron 54:677–696
Gottlieb J (2002) Parietal mechanisms of target representation. Curr Opin Neurobiol 12:134–140
Gray CM, Konig P, Engel AK, Singer W (1989) Oscillatory responses in cat visual cortex exhibit inter-columnar synchronization which reflects global stimulus properties. Nature 338:334–337
Gray CM, Engel AK, Konig P, Singer W (1990) Stimulus-dependent neuronal oscillations in cat visual cortex: receptive field properties and feature dependence. Eur J Neurosci 2:607–619
Gregoriou GG, Gotts SJ, Zhou H, Desimone R (2009) High-frequency, long-range coupling between prefrontal and visual cortex during attention. Science 324:1207–1210
Gross J, Schmitz F, Schnitzler I, Kessler K, Shapiro K, Hommel B, Schnitzler A (2004) Modulation of long-range neural synchrony reflects temporal limitations of visual attention in humans. Proc Natl Acad Sci USA 101:13050–13055
Gross J, Schnitzler A, Timmermann L, Ploner M (2007) Gamma oscillations in human primary somatosensory cortex reflect pain perception. PLoS Biol 5(5):e133
Hasenstaub A, Shu Y, Haider B, Kraushaar U, Duque A, McCormick DA (2005) Inhibitory postsynaptic potentials carry synchronized frequency information in active cortical networks. Neuron 47:423–435
Hauck M, Lorenz J, Engel AK (2007) Attention to painful stimulation enhances gamma-band activity and synchronization in human sensorimotor cortex. J Neurosci 27:9270–9277
Herculano-Houzel S, Munk MH, Neuenschwander S, Singer W (1999) Precisely synchronized oscillatory firing patterns require electroencephalographic activation. J Neurosci 19:3992–4010
Hoogenboom N, Schoffelen JM, Oostenveld R, Parkes LM, Fries P (2005) Localizing human visual gamma-band activity in frequency, time and space. Neuroimage 29:764–773
Janssen P, Shadlen MN (2005) A representation of the hazard rate of elapsed time in macaque area LIP. Nat Neurosci 8:234–241
Jin Y, O'Halloran JP, Plon L, Sandman CA, Potkin SG (2006) Alpha EEG predicts visual reaction time. Int J Neurosci 116:1035–1044
Jones MW, Wilson MA (2005) Theta rhythms coordinate hippocampal-prefrontal interactions in a spatial memory task. PLoS Biol 3(12):e402
Jutras MJ, Fries P, Buffalo EA (2009) Gamma-band synchronization in the macaque hippocampus and memory formation. J Neurosci 29:12521–12531
Kaiser J, Hertrich I, Ackermann H, Lutzenberger W (2006) Gamma-band activity over early sensory areas predicts detection of changes in audiovisual speech stimuli. Neuroimage 30:1376–1382
Kayser C, König P (2004) Stimulus locking and feature selectivity prevail in complementary frequency ranges of V1 local field potentials. Eur J Neurosci 19:485–489
Kelly SP, Lalor EC, Reilly RB, Foxe JJ (2006) Increases in alpha oscillatory power reflect an active retinotopic mechanism for distracter suppression during sustained visuospatial attention. J Neurophysiol 95:3844–3851

Khayat PS, Spekreijse H, Roelfsema PR (2006) Attention lights up new object representations before the old ones fade away. J Neurosci 26:138–142
Kopell N, Ermentrout GB, Whittington MA, Traub RD (2000) Gamma rhythms and beta rhythms have different synchronization properties. Proc Natl Acad Sci USA 97:1867–1872
Kreiter AK, Singer W (1996) Stimulus-dependent synchronization of neuronal responses in the visual cortex of the awake macaque monkey. J Neurosci 16:2381–2396
Lakatos P, Karmos G, Metha AD, Ulbert I, Schroeder CE (2008) Entrainment of neuronal oscillations as a mechanism of attentional selection. Science 320:110–113
Lakatos P, O'Connell MN, Barczak A, Mills A, Javitt DC, Schroeder CE (2009) The leading sense: supramodal control of neurophysiological context by attention. Neuron 64:419–430
Lee H, Simpson GV, Logothetis NK, Rainer G (2005) Phase locking of single neuron activity to theta oscillations during working memory in monkey extrastriate visual cortex. Neuron 45:147–156
Lin SC, Gervasoni D, Nicolelis MA (2006) Fast modulation of prefrontal cortex activity by basal forebrain noncholinergic neuronal ensembles. J Neurophysiol 96:3209–3219
Liu J, Newsome WT (2006) Local field potential in cortical area MT: stimulus tuning and behavioral correlations. J Neurosci 26:7779–7790
Luck SJ, Chelazzi L, Hillyard SA, Desimone R (1997) Neural mechanisms of spatial selective attention in areas V1, V2, and V4 of macaque visual cortex. J Neurophysiol 77:24–42
Markram H, Wang Y, Tsodyks M (1998) Differential signaling via the same axon of neocortical pyramidal neurons. Proc Natl Acad Sci USA 95:5323–5328
Markram H, Toledo-Rodriguez M, Wang Y, Gupta A, Silberberg G, Wu C (2004) Interneurons of the neocortical inhibitory system. Nat Rev Neurosci 5:793–807
Martinez-Trujillo JC, Treue S (2004) Feature-based attention increases the selectivity of population responses in primate visual cortex. Curr Biol 14:744–751
Maunsell JH, Treue S (2006) Feature-based attention in visual cortex. Trends Neurosci 29:317–322
Mishra J, Fellous JM, Sejnowski TJ (2006) Selective attention through phase relationship of excitatory and inhibitory input synchrony in a model cortical neuron. Neural Netw 19:1329–1346
Mitchell JF, Sundberg KA, Reynolds JH (2007) Differential attention-dependent response modulation across cell classes in macaque visual area V4. Neuron 55:131–141
Mitchell JF, Sundberg KA, Reynolds JH (2009) Spatial attention decorrelates intrinsic activity fluctuations in macaque area V4. Neuron 63:879–888
Monosov IE, Trageser JC, Thompson KG (2008) Measurements of simultaneously recorded spiking activity and local field potentials suggest that spatial selection emerges in the frontal eye field. Neuron 57:614–625
Montgomery SM, Buzsaki G (2007) Gamma oscillations dynamically couple hippocampal CA3 and CA1 regions during memory task performance. Proc Natl Acad Sci USA 104:14495–14500
Moran J, Desimone R (1985) Selective attention gates visual processing in the extrastriate cortex. Science 229:782–784
Munk MH, Roelfsema PR, Konig P, Engel AK, Singer W (1996) Role of reticular activation in the modulation of intracortical synchronization. Science 272:271–274
Ohara S, Crone NE, Weiss N, Lenz FA (2006) Analysis of synchrony demonstrates 'pain networks' defined by rapidly switching, task-specific, functional connectivity between pain-related cortical structures. Pain 123:244–253
Pesaran B, Pezaris JS, Sahani M, Mitra PP, Andersen RA (2002) Temporal structure in neuronal activity during working memory in macaque parietal cortex. Nat Neurosci 5:805–811
Pesaran B, Nelson MJ, Andersen RA (2008) Free choice activates a decision circuit between frontal and parietal cortex. Nature 453:406–409
Reynolds JH, Chelazzi L (2004) Attentional modulation of visual processing. Annu Rev Neurosci 27:611–647

Reynolds JH, Chelazzi L, Desimone R (1999) Competitive mechanisms subserve attention in macaque areas V2 and V4. J Neurosci 19:1736–1753
Riehle A (2005) Preparation for action: one of the key functions of motir cortex. In: Riehle A, Vaadia E (eds) Motor cortex in voluntary movements: a distributed system for distributed functions, vol 1. CDC, Boca Raton, FL, pp 213–240
Rihs TA, Michel CM, Thut G (2007) Mechanisms of selective inhibition in visual spatial attention are indexed by alpha-band EEG synchronization. Eur J Neurosci 25:603–610
Rodriguez R, Kallenbach U, Singer W, Munk MH (2004) Short- and long-term effects of cholinergic modulation on gamma oscillations and response synchronization in the visual cortex. J Neurosci 24:10369–10378
Roelfsema PR, Engel AK, König P, Singer W (1997) Visuomotor integration is associated with zero time-lag synchronization among cortical areas. Nature 385:157–161
Roelfsema PR, Tolboom M, Khayat PS (2007) Different processing phases for features, figures, and selective attention in the primary visual cortex. Neuron 56:785–792
Rudolph M, Pospischil M, Timofeev I, Destexhe A (2007) Inhibition determines membrane potential dynamics and controls action potential generation in awake and sleeping cat cortex. J Neurosci 27:5280–5290
Saalmann YB, Pigarev IN, Vidyasagar TR (2007) Neural mechanisms of visual attention: how top-down feedback highlights relevant locations. Science 316:1612–1615
Salinas E, Sejnowski TJ (2001) Correlated neuronal activity and the flow of neural information. Nat Rev Neurosci 2:539–550
Sauseng P, Klimesch W, Freunberger R, Pecherstorfer T, Hanslmayr S, Doppelmayr M (2006) Relevance of EEG alpha and theta oscillations during task switching. Exp Brain Res 170:295–301
Scherberger H, Andersen RA (2007) Target selection signals for arm reaching in the posterior parietal cortex. J Neurosci 27:2001–2012
Scherberger H, Jarvis MR, Andersen RA (2005) Cortical local field potential encodes movement intentions in the posterior parietal cortex. Neuron 46:347–354
Schoffelen JM, Oostenveld R, Fries P (2005) Neuronal coherence as a mechanism of effective corticospinal interaction. Science 308:111–113
Schroeder CE, Lakatos P (2009) Low-frequency neuronal oscillations as instruments of sensory selection. Trends Neurosci 32:9–18
Schroeder CE, Wilson DA, Radman T, Scharfman H, Lakatos P (2010) Dynamics of active sensing and perceptual selection. Curr Opin Neurobiol 20:172–176
Sederberg PB, Kahana MJ, Howard MW, Donner EJ, Madsen JR (2003) Theta and gamma oscillations during encoding predict subsequent recall. J Neurosci 23(34):10809–10814
Sederberg PB, Gauthier LV, Terushkin V, Miller JF, Barnathan JA, Kahana MJ (2006a) Oscillatory correlates of the primacy effect in episodic memory. Neuroimage 32:1422–1431
Sederberg PB, Schulze-Bonhage A, Madsen JR, Bromfield EB, McCarthy DC, Brandt A, Tully MS, Kahana MJ (2006b) Hippocampal and neocortical gamma oscillations predict memory formation in humans. Cereb Cortex 17:1190–1196
Sehatpour P, Molholm S, Schwartz TH, Mahoney JR, Mehta AD, Javitt DC, Stanton PK, Foxe JJ (2008) A human intracranial study of long-range oscillatory coherence across a frontal-occipital-hippocampal brain network during visual object processing. Proc Natl Acad Sci USA 105:4399–4404
Sejnowski TJ, Paulsen O (2006) Network oscillations: emerging computational principles. J Neurosci 26:1673–1676
Sheinberg DL, Logothetis NK (2001) Noticing familiar objects in real world scenes: the role of temporal cortical neurons in natural vision. J Neurosci 21:1340–1350
Siegel M, König P (2003) A functional gamma-band defined by stimulus-dependent synchronization in area 18 of awake behaving cats. J Neurosci 23:4251–4260
Siegel M, Donner TH, Oostenveld R, Fries P, Engel AK (2008) Neuronal synchronization along the dorsal visual pathway reflects the focus of spatial attention. Neuron 60:709–719

Simons DJ, Rensink RA (2005) Change blindness: past, present, and future. Trends Cogn Sci 9:16–20
Tallon-Baudry C, Bertrand O, Fischer C (2001) Oscillatory synchrony between human extrastriate areas during visual short-term memory maintenance. J Neurosci 21:RC177
Tallon-Baudry C, Mandon S, Freiwald WA, Kreiter AK (2004) Oscillatory synchrony in the monkey temporal lobe correlates with performance in a visual short-term memory task. Cereb Cortex 14:713–720
Taylor K, Mandon S, Freiwald WA, Kreiter AK (2005) Coherent oscillatory activity in monkey area v4 predicts successful allocation of attention. Cereb Cortex 15:1424–1437
Thut G, Nietzel A, Brandt SA, Pascual-Leone A (2006) Alpha-band electroencephalographic activity over occipital cortex indexes visuospatial attention bias and predicts visual target detection. J Neurosci 26:9494–9502
Tiesinga PH, Buia CI (2009) Spatial attention in area V4 is mediated by circuits in primary visual cortex. Neural Netw 22:1039–1054
Tiesinga PH, Sejnowski TJ (2004) Rapid temporal modulation of synchrony by competition in cortical interneuron networks. Neural Comput 16:251–275
Tiesinga P, Sejnowski TJ (2009) Cortical enlightenment: are attentional gamma oscillations driven by ING or PING? Neuron 63:727–732
Tiesinga PH, Fellous JM, Salinas E, Jose JV, Sejnowski TJ (2004) Inhibitory synchrony as a mechanism for attentional gain modulation. J Physiol Paris 98:296–314
Tiesinga P, Fellous JM, Sejnowski TJ (2008) Regulation of spike timing in visual cortical circuits. Nat Rev Neurosci 9:97–107
Tiitinen H, Sinkkonen J, Reinikainen K, Alho K, Lavikainen J, Naatanen R (1993) Selective attention enhances the auditory 40-Hz transient response in humans. Nature 364:59–60
Treue S, Martinez Trujillo JC (1999) Feature-based attention influences motion processing gain in macaque visual cortex. Nature 399:575–579
Varela F, Lachaux JP, Rodriguez E, Martinerie J (2001) The brainweb: phase synchronization and large-scale integration. Nat Rev Neurosci 2:229–239
Vida I, Bartos M, Jonas P (2006) Shunting inhibition improves robustness of gamma oscillations in hippocampal interneuron networks by homogenizing firing rates. Neuron 49:107–117
von Stein A, Chiang C, König P (2000) Top-down processing mediated by interareal synchronization. Proc Natl Acad Sci USA 97:14748–14753
Wannig A, Rodriguez V, Freiwald WA (2007) Attention to surfaces modulates motion processing in extrastriate area MT. Neuron 54:639–651
Womelsdorf T, Fries P (2007) The role of neuronal synchronization in selective attention. Curr Opin Neurobiol 17:154–160
Womelsdorf T, Fries P, Mitra PP, Desimone R (2006) Gamma-band synchronization in visual cortex predicts speed of change detection. Nature 439:733–736
Womelsdorf T, Schoffelen JM, Oostenveld R, Singer W, Desimone R, Engel AK, Fries P (2007) Modulation of neuronal interactions through neuronal synchronization. Science 316:1609–1612
Worden MS, Foxe JJ, Wang N, Simpson GV (2000) Anticipatory biasing of visuospatial attention indexed by retinotopically specific alpha-band electroencephalography increases over occipital cortex. J Neurosci 20:RC63
Wrobel A, Ghazaryan A, Bekisz M, Bogdan W, Kaminski J (2007) Two streams of attention-dependent beta activity in the striate recipient zone of cat's lateral posterior–pulvinar complex. J Neurosci 27:2230–2240
Wyart V, Tallon-Baudry C (2008) Neural dissociation between visual awareness and spatial attention. J Neurosci 28:2667–2679
Yamagishi N, Callan DE, Goda N, Anderson SJ, Yoshida Y, Kawato M (2003) Attentional modulation of oscillatory activity in human visual cortex. Neuroimage 20:98–113

Studying Consciousness Using Direct Recording from Single Neurons in the Human Brain

Moran Cerf and Michael Mackay

Abstract Consciousness represents a unity of experience derived from limitless possibilities and fierce competition for dominance within our brain. This unity reflects the compromise of an internal representation of the world that shapes our perception of the environment and guides the choices we make in interacting with it. Here we investigate the relationship between the external environment, our internal representation of it, and volition, using direct recording from single neurons in the brains of 12 patients with pharmacologically intractable epilepsy. In two novel experiments we pit the world within against the world outside, and separate the decision making processes within the brain from the sensation of volition. We show that subjects are capable of overriding external sensory input with internal imagery, and can directly control the firing rate of individual neurons in the medial temporal lobe. Additionally, we show that decision can be predicted before the conscious perception of volition, but that feeding this decision forward and bypassing volition leads to a gradual loss of predictive ability as the neurons alter their activity in response. We propose a working definition for consciousness based on these results, suggesting that consciousness involves our memory and experience of events and ties these together in a working narrative.

1 Introduction

Studies in psychology separate two entities competing in our brain for dominance. One is the "experiencing I," characterized as "living in the present," intuitive, and effortless. This entity, commonly termed "system 1," is automatic, opaque, skilled

M. Cerf (✉)
Computation and Neural Systems, California Institute of Technology, Pasadena, CA 91125, USA
and
Department of Neurosurgery, University of California Los Angeles, Los Angeles, CA 90095, USA
and
School of Business, New York University, New York, NY 10012, USA
e-mail: moran@morancerf.com

S. Dehaene and Y. Christen (eds.), *Characterizing Consciousness: From Cognition to the Clinic?* Research and Perspectives in Neurosciences,
DOI 10.1007/978-3-642-18015-6_8, © Springer-Verlag Berlin Heidelberg 2011

and associative. The other entity, "system 2," is the "remembering I," which is deductive, slow, self-aware and top–down controlled (Sloman 1996). The latter is the one we commonly interact with consciously. When we are about to choose a vacation, for instance, we put the two to competition: the experiencing system wants to go to a new place and generate more experiences to live by, while the remembering one wants to choose the best memory from a list of existing memories – a memory that was already proven to be positive.

These systems reflect a fundamental neurological separation of the outside, sensory world from its internal representation in our brain (Cerf et al. 2010). The ability to separate that which is internally generated from that which is externally generated is arguably what underpins our identity (Frith 1987). Neuroscience often strives for a view of the world where a representation of the environment exists directly within our brain – a "what you see is what you get" type of phenomenon. The outside world is then represented in our brain as is. But in life we know that we can shape the information in our mind in multiple ways: information from the environment can be stored in our brain in a less informative set of values than the actual experience (Redelmeier and Kahneman 1996), information retrieved from our memory can be reconsolidated with changes (Schiller et al. 2010), and information can be manipulated by our senses (Cerf et al. 2010). Thus we find a blurring of the lines between the internal representation of the world and its external manifestation. This separation is conceptual but also exists neurologically in the form of different regions within the brain.

While our brain can be thought of as a coalition of various sub-parts, we normally think of our personality as a coherent entity. That said, we know that these sub-parts occasionally vie for dominance within the brain (Kreiman et al. 2002). Studies such as the ones conducted with split-brain patients – patients whose right and left hemispheres were separated in surgery for clinical reasons – explicitly demonstrate this competition because the two hemispheres can be given conflicting tasks. The two halves of the brain, when separated, act as two individual entities, each controlling their respective half of the body and responding to their half of the sensory world (Gazzaniga 2005). Such phenomena could happen in all our brains with binocular rivalry tasks (Leopold and Logothetis 1996), where different images which are presented to the two eyes yield conflicting percepts. However, in normal subjects this competition is resolved moment to moment to give a single conscious perception, despite its fluctuant nature (Levelt 1965).

The actions we project to the world are the sum of internal ruminations and deliberations, which are commonly aggregated into a single decision. Once a decision is made, there then occurs one of the most debated phenomena in philosophy of mind: volition. Though impossible to discern in its metaphysical guise, modern neuroscience has opened the fundamental question of "free will" up to scientific investigation by looking for the signature of decision and motor initiation in the brain. To do so a range of techniques has been used: from direct electrical stimulation to cutting edge neuroimaging and animal neurophysiology. Evidence of a volitional signal emerged with the recording of the *bereitschaftspotential* (Kornhuber and Deecke 1965; Libet et al. 1983; Libet 1999), which was a rising level of activity in motor cortex preceding voluntary muscle movement. Scalp EEG and MEG experiments with humans then revealed changes in neural activity that preceded awareness of volitional state by hundreds of milliseconds (Haggard 2008).

We now know that there are multiple stages to motor initiation: the rise of a "readiness potential," the conscious urge to act, and the action itself (Fried et al. 2011). While the initial change in neural activity can precede the conscious urge by as much as seconds, the urge and the action are typically separated by 100–200 ms.

These separate stages allowed us to probe two intriguing aspects of volition: firstly, can a volitional signal be decoded on a single neuron level, and secondly, what happens when the brain is made aware of its decision before the conscious observer is? Furthermore, we were interested in seeing what changes could be incurred in the brain following direct tampering with its ability to control the flow of information.

To address these questions, we designed two experiments that offered human subjects the ability to alter the activity of neurons in their brains while getting direct feedback on this activity. Here we explore how and where the interaction between our external and internal worlds occurs in the brain by directly pitting the world within and the world outside against each other. In doing so, we directly test the ability of subjects to alter the dominance of brain regions and show evidence for the ability of humans to shape the way information is perceived in their brains.

2 Experiments

The subjects in these experiments were 12 patients with pharmacologically intractable epilepsy who were undergoing brain surgery to localize the seizure focus for possible surgical resection (Fried et al. 1997). During this procedure, the patients were implanted with intracranial electrodes accessing regions in the medial temporal lobe (hippocampus, amygdala, entorhinal cortex and parahippocampal cortex). After surgery, subjects participated in a series of experiments. In the first experiment, they were instructed to play a game in which they controlled the display of two superimposed images via the firing activity of four medial temporal lobe (MTL) neurons in the brain (Fig. 1a). In a prior screening session, we had identified four different neurons that responded selectively to four different images by increasing their firing rate when the corresponding image was presented. Each trial started with a two-second display of one of these four images (the target). Subjects next saw an overlaid hybrid image consisting of the target and one of the three remaining images (the distractor) and were told to enhance the target ("fade in") by focusing their thoughts on it. The initial visibility of both images was 50% and was adjusted every 100 ms by feeding the firing rates of four MTL neurons into a real-time decoder that could change the visibility ratios until either the target was fully visible ("success"), the distractor was fully visible ("failure"), or until 10 s had passed ("timeout").

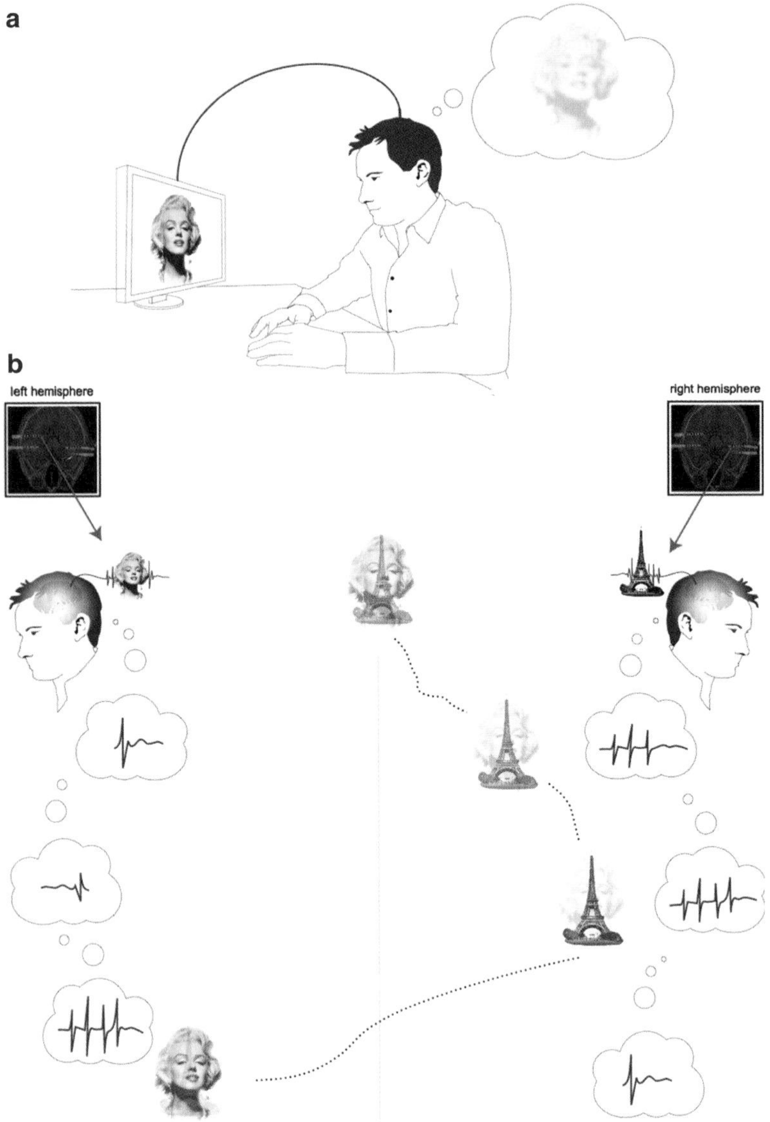

Fig. 1 (**a**) Recording from intracranial electrodes, neurons are identified that respond to a specific concept. In this instance a cell responsive to the image of Marilyn Monroe was found. This cell increases its firing rate to the image or thought of Monroe. (**b**) This cell is then pitted against one found to represent the Eiffel tower. The two images are superimposed and the subject is asked to bring the image of Monroe to maximum visibility. The visibility of the image is controlled by real-time decoding of the activity of each neuron relative to the other neuron and its own baseline. In this example, we show a case where the subject initially begins to fail the experiment – the firing of the Eiffel neuron increases and the visibility of the tower increases, creating negative feedback. However, the subject is able to exert control and, by concentrating on the internal thought of Monroe, is able to override this sensory input and increase the firing rate of the Monroe neuron and decrease that of the Eiffel neuron, bringing the image of Monroe to visibility. The scans show the location of the respective electrodes within the brain

The subjects manipulated the visibility of the hybrid images by any cognitive strategy of their choosing. Six of 12 subjects reported in a follow-up interview that they focused on the concept represented by the target picture or closely allied associations. As they performed the task, we showed that subjects could control their own brains as though they were controlled by an external device (Cerf et al. 2010).

Occasionally the subject nearly failed in the task (Fig. 1b). While they were told to focus their thoughts on a particular concept (say, Marilyn Monroe), which was pitted against a different concept (say, the Eiffel tower), the distractor gradually became more dominant on the screen, nearly leading to a failure by becoming fully visible. However, just when the subject was about to fail the trial, he/she was able occasionally to summon the thought of the target concept into his/her mind, making the neuron or coalition of neurons corresponding to Marilyn Monroe engage in activity and shift the hybrid image back to the target. This finding implies that, while visual feedback showing an image of the Eiffel tower was penetrating the brain, the subject was able to override this information with the mental imagery of Monroe. Thus, this experiment has a deeper and more profound result: imagery can override vision at the level of the MTL, or as we choose to put it, idealism trumps realism. The world in the subject's brain can override the outside world.

Practically, this is equivalent to seeing a cup of tea with your eyes, but choosing, in Kantian terminology, to wear different glasses, such that the image of the cup is turned inside our mind to a picture of a flower. Our "flower" neuron would fire and we would actually "see" a flower, although the photons hitting our retina would reflect a cup of tea. The same analogy can be used for any stimuli: hearing a word differently that what it was when a person swore at you; choosing to ignore pain; or choosing to not let any external stimulation penetrate when you focus your mind within. We can synthesize an internal world to override the one given to us by the senses.

If this merging of the external and internal worlds is constantly happening, then one can ask what our conscious experience actually represents. From a purely sensory basis, we would want this information to be the most accurate representation of reality that our sensory systems are capable of generating. However, we experience, learn, and create an internal representation of the world, describing how we expect it to behave and how we interact with it. Thus the phenomenal reality we perceive is a blend of these two worlds – the objective, external world, shaped by our subjective, internal world – to create what is effectively a learned representation of reality.

In this study, we show evidence of humans' ability to alter the way information flows in the brain, using information from the visual system. Practically, this can be extrapolated to any procedure of experiencing the outside world. For any means of altering perceptual information, it is likely that our "remembering I" performs these kinds of modifications to our sensory input constantly, thus underlying our well-described predisposition to perceive the world we expect to perceive (Merleau-Ponty 1996).

An important note to address in this experiment is the question of what happens at the neuronal level. To answer that question, we measured the change in activity of neurons in the brain pitted together in this competition. A success in each trial in which two neurons are pitted against each other can be the result of three possibilities: the neuron selective for the target (target neuron) increasing its firing above its baseline, the distractor neuron in this competition decreasing its firing below its baseline, or the two neurons working together, with target rising and distractor declining. Testing this point directly shows that the third case is the most common. In over 80% of the trials, two neurons, often separated in two hemispheres and distant from each other, ended up working together to perform the task, with one neuron significantly increasing its firing rate while the other decreasing its own. This sort of entanglement between two neurons is profoundly surprising. The two neurons, along with the subject, had no knowledge prior to the experiment which two images would end up being placed in competition. The images were selected based solely on the morning's screening. Moreover, each trial pitted together a different pair of images not known to the subject in advance. However, as soon as the trial began and the subject realized which two images were competing, they immediately controlled one neuron to go up while the other went down, channelling information far into two different locations rapidly.

A question that comes to mind is what coordinates this activity such that both neurons simultaneously alter their activity to work together to perform the task so rapidly and profoundly? We know from testing sets of neurons all firing for the same concept that the effect is not purely working on an individual neuron but on a small coalition of neurons all responding to the same concept (Cerf et al. 2010). We also know that this effect is very sparse (Waydo et al. 2006). Only the Monroe and Eiffel neurons will alter their behavior and not a larger region of the brain (Cerf et al. 2010). Our results show that neurons adjacent to the Monroe/Eiffel neurons will not change their firing patterns in this trial; only the two relevant clusters of neurons will. More specifically, in a following trial, we could pit Monroe with a third image, say, an image of Johnny Cash. In this trial, Monroe neurons would go up while Cash neurons go down, but the Eiffel tower neurons from the trial before would remain untouched. If in a following trial Cash and the Eiffel tower would compete, the Monroe neurons will remain still while the Eiffel/Cash neurons go up/down or down/up, based on the target of the trial.

These changes happen rapidly and accurately, and the specificity suggests that this effect is not simply the result of lateral inhibition. These results fit with those seen in neurophysiological studies of attention in primates (Reynolds et al. 1999), in that the change in response is seen after a delay and it increases if the neuron represents the attended image but decreases if that neuron's target is not attended to. This finding suggests that the coordinators of such changes in response may be attentive processes, which makes sense in the context of the subject's reported cognitive strategies.

Finally, it is important to note that the characterization, properties and behavior of these neurons seem very similar to those of place cells in rodents. If one assumes that thoughts in "concepts space" are as relevant to humans as locations in "place fields," and that humans practically navigate their lives within such a concept space as rodents navigate in a maze, then one can draw this analogy between the properties of the two and suggest that the MTL is the seat of the dominant entity that we use to navigate by in our minds. Humans travel inside their thoughts as rodents navigate the world in a maze. Our two-images competitions show strong analogy to the results of studies showing experiments in which rodents were given a two-way maze to choose from and the typical firings of two single place cells were pitted against each other in individual trials (O'Keefe and Nadel 1978; Wilson and McNaughton 1993; Eichenbaum 1996; Moser et al. 2008).

As the first experiment demonstrated the subjects' ability to increase or decrease the activity of single neurons in their brains voluntarily using their sheer will, we wanted additionally to consider what changes could occur in the same neurons without conscious control but based on feedback from the environment. To do so, we designed an experiment that separated the subjects' desire to act from the neuronal mechanisms that precipitated the feeling of volition. Following a prior experiment, which showed that it is possible to decode the activity of single neurons in subjects' brains prior to their volition to act (Fried et al. 2011), we designed an additional on-line experiment that interacted with these neurons prior to the subjects' conscious knowledge of the act.

The subjects were faced with two red buttons and were asked to press either the left or right one with their hands, in their own time. Each trial started with the subject having to wait for three seconds until a "get ready" sign disappeared. Once the signal disappeared, the subjects had as much time as they wanted to feel the urge to press a button before choosing which one to press, and as they did so the button was "lit" (turned on). The subjects then waited 3 s for the light to turn off before another trial began. Occasionally, the button was turned on without the subject pressing it. The subjects were told that this sometimes happens and were asked to wait for the button to be turned off again and for the following trial to start. Subjects were not encouraged to press the button quickly or slowly but were specifically told to wait until they felt the urge as need be. They were only instructed not to press the button when the light is on.

The first 50 trials were used to train a decoder on identifying sets of neurons that fired prior to the subject's urge to move. We identified neurons that fired between 1,500 and 200 ms prior to the subject's movement and between 1,000 and 100 ms prior to the subject's reported "will" to initiate a movement. Once the computer identified these neurons, or a classifier passed the threshold of 75% accurate predictions of the following button press at least 300 ms in advance, we began the active part of the experiment. While the subject still performed the same task, we altered the mechanism that turned the button on. Contrary to the original trials, in which the choice to turn the button on was either based on a press of the button or random timing, in the second phase we turned the button on either randomly (30% of the trials) or based on a decoding of the subject's will to press a button (Fig. 2).

As we decoded the subject's will to press a specific button prior to their conscious perception of this will, we generated two more conditions whereby we turned the predicted button on at different times. Either we intentionally waited

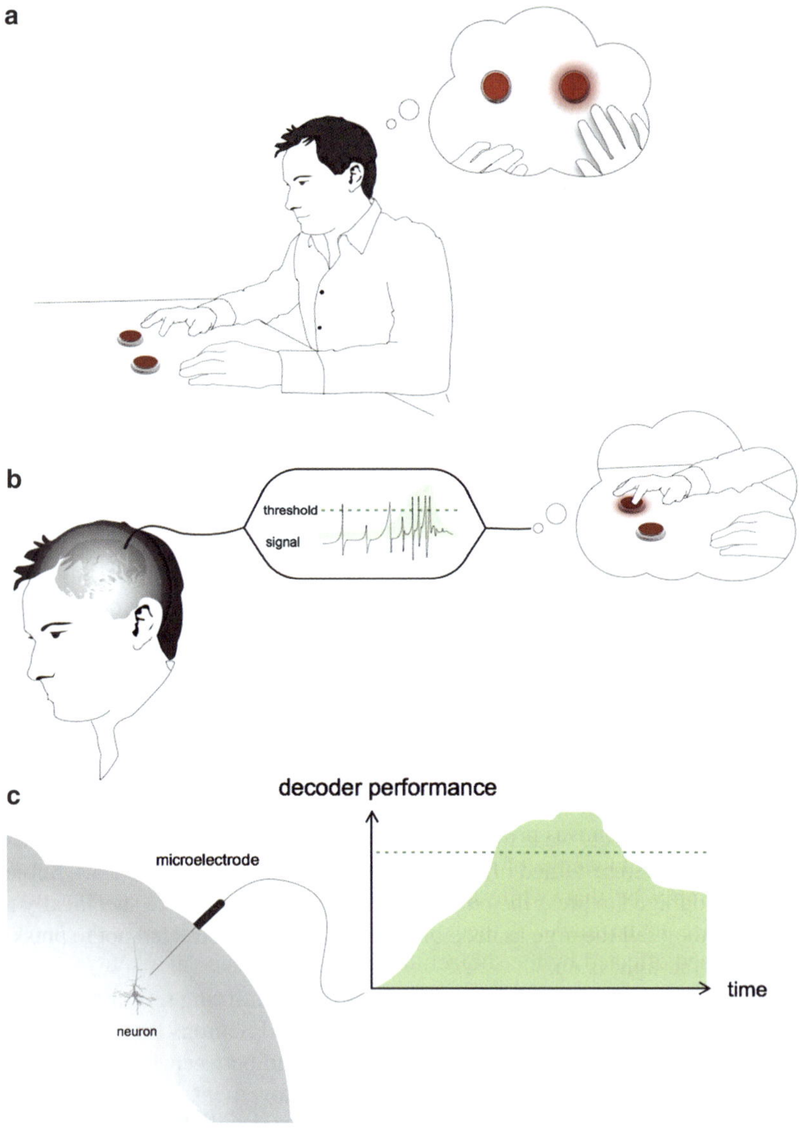

Fig. 2 (**a**) The subject is faced with two buttons. At the start of a trial, he/she is allowed to choose at will which button to press when he/she feels the urge to do so. On pressing the button, it lights up. In some trials a button lights up prior to the subject pushing it, based on the decoding of the urge to press the button. (**b**) Selected neurons show an increasing rate of activity until a threshold is reached at which the subject feels the urge to move, but before motor initiation. (**c**) For the initial trials, this activity of neurons predicting volition was decoded until capable of accurately decoding the subjects' choice prior to the conscious urge to move. Once the threshold of 75% accuracy was reached, we inserted a condition in which we used this information to light the predicted button up prior to the subject feeling of urge to press it. After a number of trials, the success of the decoder started to reduce as the neuron altered its activity, presumably in response to being decoded

500 ms (the presumed time it takes the subject to become aware of his/her will to press the button) and then turned the button on, or immediately turned the button on, such that the subject was unaware of his/her brain's presumed will to press the button. The third condition created a unique situation in which brain activity gave rise to an action in the environment without the subject's conscious awareness that he/she had triggered it.

From the subject's point of view, the "random" trials and the predicted activation trials were identical. In both cases, the button was turned on without the subject feeling that he/she had any effect on it. The difference, though, is in the "neuronal" view of the experiment. In this case, the neurons (to anthropomorphize them) did know that they were active and were signalling their "will" to press the button. The feedback from the environment confirmed to these neurons that what they wanted was done; however, the internal feedback said that the result was not triggered by the subject.

Our results show that, on the single neuron level, looking at the decoding ability of an individual neuron that is identified as a key module in giving rise to volition, the neurons reached a plateau of maximal decoding within the first trials and were being decoded in real time during the following trials. However, we saw a gradual decrease in the performance of these neurons as the experiment progressed. Simply put, the neuron learned that it was being decoded and that the decoding interfered with the task at hand and, within a few trials, learned to not allow for the decoding, all without the subjects' knowledge. The subjects showed no conscious knowledge of the shift in decoding or in the ability to affect the trials that were based on the neuron's activity, but we saw that the decoder's performance was dropping based on the success of the decoder to interfere with the neuronal activity. Practically, the subjects just claimed to have no desire to press the button for a few trials (the trials within which the button was pressed because of their unconscious will), after which they regained the knowledge of their will and again felt the desire to act.

It may simply be that, while these neurons initially signal that they want a specific choice to be made, if there is repeatedly a lack of feedback that their "will" has been effected by the subject then they may reduce in activity.

3 Discussion

In the first experiment, we demonstrated the subjects' ability to up- and down-regulate the activity of neurons voluntarily using their thoughts. In the second experiment, we showed neurons changing their behavior based on a disparity in feedback from the internal and external worlds. Both experiments demonstrate direct access to the activity of single neurons in the brain without changing the entire sequence of activity of neighboring neurons. Moreover, both experiments show evidence of a change in network activity based on the task, altered either by attention or by the effect of neuronal response on the environment. However, while the first experiment happens with conscious and direct control arriving from the subject, the second one bypasses the subject's awareness. This raises a question:

when neuronal alteration of behavior occurs in the brain due to feedback, who is getting the feedback and exerting the change?

It might appear as if there are two people involved in these experiments. One is the subject's mind, instructing the brain to think of a concept, while the other is the one that is acting out the mind's desire, namely, the nerve cells in the MTL that up- and down-regulate their activity accordingly. A fundamental peculiarity of the mind–body nexus is apparent here: we do not think in terms of neuronal firing rates. Our thoughts, desires and feelings are metaphysical entities. We have no knowledge of the mechanisms that interact inside our brains when we act. Yet the thought of a concept such as Marilyn Monroe translates into a particular pattern of neuronal activity when we elicit her image in our minds (Kreiman et al. 2000). Events in the phenomenal mind find their parallel in the material brain. We have to wonder, however, if it is our conscious desire to "see" Monroe that drives the change in neuronal firing or the change in neuronal firing that generates our desire. Who is the "puppet" and who is the "puppeteer?"

Practically, it is the brain that fulfils both roles; the same brain is learning about the task and performing it. The "puppet" is also the "puppeteer" and most likely represents a circular route of causality. There is a singular level of neuronal coalitions that represent conscious perceptions that is subject to incoming sensory information and is then appraised relative to our internal world and desires, and which then feeds back and modifies the activity of these same neurons. Incoming information then simply feeds into this eternal loop.

One can ask then, when brains communicate, either with the outside world or, for that matter, with themselves, through internal ruminants or via imagery, what gives us the illusion that one person is "listening" while the other is "performing?" Why, when speaking to someone, despite both parties hearing the speech simultaneously, do we somehow feel that these thoughts are owned by us more than they are owned by others (Stephens et al. 2010)?

It is worth noting that the brain separates the internal world from the external all the time. In doing so, it attaches a profound sense of ownership to conscious sensations that it determines as internally generated, such as our internal voice or the sensation of moving one's body. Indeed, one of the most devastating psychiatric disorders, schizophrenia, can be explained in terms of the brain's inability to correctly separate out these two streams of sensory information (Frith 1992). Thus the internal voice is misattributed as an external sensation. At the same time, a fundamental loss of sense of self may underlie thought interference, in which a person believes his/her thoughts are open to being read, or withdrawn, or inserted by another party; the person loses the profound sense of ownership we have of our own thoughts (Nelson et al. 2009).

In this context, the experiment in volition provides a remarkable window on the processes that underlie this attribution of ownership to sensory information about external events. We have a decoder that can accurately predict a decision before the conscious perception of will, but also before a motor command is generated. We also know that the subject feels no ownership of the button being lit up if it does so before they become aware of the will to press it (roughly 100–500 ms after the neuronal activity the decoder reads). Thus we have bisected the process at a point where what has yet to happen is crucial to applying a sense of ownership to the button lighting up, but parts of the brain have already prepared and effectively selected the decision to come. This finding allows us to look at how the brain responds to such a contradiction but also potentially allows us, in the future, to dissect out what contributes to volition and the sense of ownership.

This relationship between the internal world and the external world is important in determining the content of consciousness. With that in mind, one might ask if we can access a person's consciousness without him/her being involved whatsoever. In effect, the advertising industry does this all the time: making us buy the soup that they want without noticing that we wanted it (we do not know any more if we like the taste of Coke; we just know it as something we like). In the same way, we were able to implant in our subjects the occasional firing of a neuron either with the aid of their conscious volition or with the aid of the prior mechanisms that gave rise to that sense of volition, even without an active feeling of agency; without generating the thought in their mind.

The reason for the discrepancy between the conscious feeling of agency in the first experiment and the unconscious one in the second could be attributed to memory. Memories are an essential second component of consciousness. Memories tie together the set of stories we have into a narrative. That is because the thought requires an additional component on top of the activity of these neurons: it requires a narrative to be built from it. The sensory world we are immersed in is translated into neuronal representations, and these representations are dependent on past experiences, our internal world and its narrative. The essence of consciousness is the material that selects the one percept from the many that rises to our awareness. To that effect, our experiment is a demonstration of humans' ability to control what is accessible to their minds. Our subjects could practically choose to not see things that were visible to their eyes and see things that were not, purely based on thoughts. In doing so, they could actively select which patterns to encode in their memory and which not to.

This is almost a Buddhist point of view: the world in front of our eyes is just a suggestion whereas the world inside our brain is ours to create. Put differently, while some aspects of the world seem to occur to us, much of it is ours to construct. The person cursing you on the street because you cut him off driving is sending an auditory signal that will most certainly be processed by your ears and triggering activity in your auditory system, but whether you let it alter your consciousness or not is up to you and the narrative you are building. When we walk in the streets and see three little kids walking towards us, the photons from the environment are hitting our eyes simultaneously. But only a complex circuit in one's brain can lead to "noticing" one's son on the left first and not the other two children to his right. Presumably, for the same visual input a different parent will notice the girl in the center first and will give rise to a different neuronal representation despite an initial visual response that is nearly identical.

Additionally, these two experiments touch on the question of what parts of our brain we are capable of exerting conscious control over. While subjects in our experiment exerted control over brain regions such as the hippocampus and parahippocampal cortex, the same effect was not seen in great magnitude in other regions such as the amygdala or entorhinal cortex (Cerf et al. 2010). The latter showed an average performance level of 20% compared to nearly 70% in the hippocampus and 69% in the parahippocampal cortex. The ability or lack of ability to control our brains in regions such as the amygdala remains open for further investigation. The reasons behind the lack of control in some regions are not clear yet. However, one can propose that this lack of control is tied to our general conception of hardship in controlling our emotions. With the amygdala being implicated as a primary seat of emotions, and our experiments showing a decreased level of ability to consciously control the amygdala neurons voluntarily, one can propose that this is evidence of the lack of ability to regulate emotions in some brain regions directly. While adult humans are able to regulate their emotions to some extent (unlike young humans or most animals which – for better and for worst – seem to be in the moment at any given time) we do not seem to show the same level of control over those as we show over our memories and decisions. We can summon a thought to our mind rather easily, but cannot control emotions as easily. We cannot choose when to feel things and exactly how and what to feel. Emotions somehow seem to dawn on us. They capture us and we steer our boat through them. At the extreme end of the spectrum of control, we have those who suffer from bipolar affective disorder, whose emotions are inherently unstable, uncontrollable, and excessively powerful.

As the question we posed in the beginning addressed not only the experimental results from studies of single neurons but also their relevance to the study of consciousness, we want to end by suggesting a working definition of what consciousness is and why it is something that we can probe best using studies of with single neurons in humans. Consciousness is the narrative that we compose about our actions. It is the story we make from the series of occurrences and events in our brain, in the form of past memories, current feelings and thoughts, simulations of the future, and sets of influences from the environment, be it via the five senses, internal rumination, or random guided processes that we attribute to our own free will. The major purpose of consciousness, therefore, is to communicate this story, this narrative, to as many equal brains as possible. The way to immortalize ourselves is by creating multiple instances of our existence, as accurately as possible, in as many brains as possible.

The idea of neurons encoding us (like the Marilyn Monroe neurons) could be, in that same point of view we remarked on beforehand, the ultimate way to sustain our existence. Having children and creating a new generation is such a guaranteed imprint of ourselves in an additional brain. Interaction with other humans and living organisms and accurate communication are the methods by which we ensure that our imprints and our experiences, as they are imprinted in other brains, are as accurate as possible. It is how we evolve ourselves and make sure we keep improving the source of those imprints in such a way that will allow us to increase the possibilities of future encoding of ourselves in more brains. The smarter we become and the better communicators we become, the more likely we are to generate more such instances and more accurate ones.

We walk our story, talk our story and even live it when we are asleep. Every occurrence, even the ones we are not aware of, or the ones we prefer to not have experienced, shapes this narrative and offers a reflection on our existence that could lead to improvement in our behavior. Consciousness, therefore, requires knowledge of the source of its perceptions and for those perceptions to be consistent in the context of its ongoing narrative; it wants to always have an answer to what is happening regarding itself. It is the reason we always come up with an explanation for situations and the reason we try to constantly regulate our emotions and find it hard to accept that they dawn on us. It is the reason we suffer whenever we feel that our free will is in jeopardy, and the reason we conduct experiments that teach us how to control our own neurons. It is a step towards understanding the mechanisms to get closer to our story and make sure we can back it up after our bodies cease to exist. It is the main reason humans are so baffled by consciousness and the reason they spend so much time and energy using their brain to reflect and ponder about itself, study its own behavior, write papers and draw conclusions about itself. It is a quest for the nature of oneself.

Acknowledgment We wish to thank the patients for their participation in these studies; Daniela Schiller, Liat Maoz and Udi Pladott for help with the manuscript, and Maria Moon for help with the illustrations; Sonia Le Cornec and Yves Christen from the IPSEN foundation for support of this work. This paper is based on work conducted with Profs. Itzhak Fried and Christof Koch (Cerf et al. 2010).

References

Cerf M, Thiruvengadam N, Mormann F, Kraskov A, Quiroga RQ, Koch C, Fried I (2010) On-line, voluntary control of human temporal lobe neurons. Nature 467(7319):1104–1108

Eichenbaum H (1996) Is the rodent hippocampus just for place? Curr Opin Neurobiol 6(2):187–195

Fried I, MacDonald KA, Wilson CL (1997) Single neuron activity in human hippocampus and amygdala during recognition of faces and objects. Neuron 18(5):753–765

Fried I, Mukamel R, Kreiman G (2011) Internally generated preactivation of single neurons in human medial frontal cortex predicts volition. Neuron 69(3):548–562

Frith C (1987) The positive and negative symptoms of schizophrenia reflect impairments in the perception and initiation of action. Psychol Med 17(03):631–648

Frith C (1992) The cognitive neuropsychology of schizophrenia. Lawrence Erlbaum Associates, Hove, UK

Gazzaniga MS (2005) Forty-five years of split-brain research and still going strong. Nat Rev Neurosci 6(8):653–659

Haggard P (2008) Human volition: towards a neuroscience of will. Nat Rev Neurosci 9(12):934–946

Kornhuber HH, Deecke L (1965) Hirnpotentialänderungen bei Willkürbewegungen und passiven Bewegungen des Menschen: Bereitschaftspotential und reafferente Potentiale. Pflügers Archiv Eur J Physiol 284(1):1–17

Kreiman G, Koch C, Fried I (2000) Imagery neurons in the human brain. Nature 408(6810): 357–361

Kreiman G, Fried I, Koch C (2002) Single-neuron correlates of subjective vision in the human medial temporal lobe. Proc Natl Acad Sci USA 99(12):8378

Leopold DA, Logothetis NK (1996) Activity changes in early visual cortex reflect monkeys' percepts during binocular rivalry. Nature 379(6565):549–553

Levelt WJM (1965) On binocular rivalry. Institute for Perception RCO-TNO, Soesterberg, Netherlands

Libet B (1999) Do we have free will? J Consciousness Stud 6(8–9):47–57

Libet B, Gleason CA, Wright EW, Pearl DK (1983) Time of conscious intention to act in relation to onset of cerebral activity (readiness-potential): the unconscious initiation of a freely voluntary act. Brain 106(3):623

Merleau-Ponty M (1996) Phenomenology of perception (trad: C. Smith). Motilal Banarsidass, Delhi

Moser EI, Kropff E, Moser MB (2008) Place cells, grid cells, and the brain's spatial representation system. Neuroscience 31(1):69

Nelson B, Fornito A, Harrison BJ, Yücel M, Sass LA, Yung AR, Thompson A, Wood SJ, Pantelis C, McGorry PD (2009) A disturbed sense of self in the psychosis prodrome: linking phenomenology and neurobiology. Neurosci Biobehav Rev 33(6):807–817

O'Keefe J, Nadel L (1978) The hippocampus as a cognitive map. Clarendon, Oxford

Redelmeier DA, Kahneman D (1996) Patients' memories of painful medical treatments: real-time and retrospective evaluations of two minimally invasive procedures. Pain 66(1):3–8

Reynolds JH, Chelazzi L et al (1999) Competitive mechanisms subserve attention in macaque areas V2 and V4. J Neurosci 19(5):1736

Schiller D, Monfils MH, Raio CM, Johnson DC, Ledoux JE, Phelps EA (2010) Preventing the return of fear in humans using reconsolidation update mechanisms. Nature 463(7277):49–53

Sloman SA (1996) The empirical case for two systems of reasoning. Psychol Bull 119:1.3–22

Stephens GJ, Silbert LJ et al (2010) Speaker-listener neural coupling underlies successful communication. Proc Natl Acad Sci USA 107(32):14425

Waydo S, Kraskov A, Quian Quiroga R, Fried I, Koch C (2006) Parse representation in the human medial temporal lobe. J Neurosci 26(40):10232

Wilson MA, McNaughton BL (1993) Dynamics of the hippocampal ensemble code for space. Science 261(5124):1055

Intrinsic Activity and Consciousness

Marcus E. Raichle

Abstract Traditionally, studies of brain function have focused on task-evoked responses. By their very nature, such experiments tacitly encourage a reflexive view of brain function. While such an approach has been remarkably productive, it ignores the alternative possibility: that brain functions are mainly intrinsic, involving information processing for interpreting, responding to and predicting environmental demands. I shall argue that the latter view best captures the essence of brain function, a position that accords well with the allocation of the brain's energy resources. Understanding intrinsic activity will require integrating knowledge from cognitive, systems, cellular and molecular neuroscience. Ultimately, understanding conscious awareness will require an appreciation of the fundamental role played by intrinsic activity in brain function.

> "*Whilst part of what we perceive comes through our senses from the object before us, another part (and it may be the larger part) always comes out of our own head*"
>
> William James (1890)

This prescient comment by William James, to be found in Volume 2 (page 103) of his monumental work, Principles of Psychology (James 1890), captures the essence of a debate ongoing in the nineteenth century and possibly earlier surrounding two views of brain function. One view, pioneered by the work of Sir Charles Sherrington (Sherrington 1906), posits that the brain is primarily reflexive, driven by the momentary demands of the environment. The other view is that the brain's operations are mainly intrinsic, involving the acquisition and maintenance of information for interpreting, responding to and even predicting environmental demands, a view espoused by one of Sherrington's disciples, T. Graham Brown (Brown 1914) (for more recent perspectives see Llinas 2001; Yuste et al. 2005).

M.E. Raichle (✉)
Departments of Radiology, Neurology, Neurobiology and Biomedical Engineering, Washington University School of Medicine, 4525 Scott Avenue, Room 2116, St Louis, MO 63110, USA
e-mail: marc@npg.wustl.edu

S. Dehaene and Y. Christen (eds.), *Characterizing Consciousness: From Cognition to the Clinic?* Research and Perspectives in Neurosciences,
DOI 10.1007/978-3-642-18015-6_9, © Springer-Verlag Berlin Heidelberg 2011

The view that the brain is primarily reflexive has motivated most neuroscience research, including that with functional neuroimaging. This result is not surprising because experiments designed to measure brain responses to controlled stimuli and carefully designed tasks can be rigorously controlled, whereas evaluating the behavioral relevance of intrinsic activity (i.e., ongoing neural and metabolic activity that is not directly associated with subjects' performance of a task) can be an elusive enterprise. Unfortunately, the success of studying evoked activity has caused us to lose sight of the possibility that our experiments reveal only a small fraction of the actual functional activity performed by our brain.

Two challenges face a consideration of the view that the brain's operations are mainly intrinsic. First, how do we adjudicate the merits of such a claim? The answer comes primarily from a consideration of the considerable cost of running the brain, most of which is devoted to its ongoing, internal functional activity. And, second, if the claim is correct, how do we unlock the mysteries of intrinsic activity? The answer will come from a serious consideration of multiple levels of inquiry, ranging from cognitive and systems neuroscience to cell biology and metabolism. A much expanded version of what is to follow can be found in an article entitled, "Two views of brain function" (Raichle 2010).

1 Adjudicating the Merits of Intrinsic Activity

1.1 Cost

One of the most persuasive arguments for the importance of intrinsic activity emerges from a consideration of its relative cost in terms of brain energy consumption. In the average adult human, the brain represents about 2% of the total body weight, yet it accounts for 20% of all the energy consumed (Clarke and Sokoloff 1999), 10 times that predicted by its weight alone.

Relative to this very high rate of ongoing energy consumption in the resting state, the additional energy consumption associated with changes in brain activity is remarkably small, often less than 5% (Raichle and Mintun 2006). From these data it is clear that the brain's enormous energy consumption is little affected by task performance, an observation first made more than 50 years ago by Louis Sokoloff, Seymour Kety and their colleagues (Sokoloff et al. 1955) but rarely cited.

What is the nature of this ongoing intrinsic activity that commands such a large amount of the brain's energy resources? Assessments of brain energy metabolism using a variety of approaches (Sibson et al. 1997, 1998; Ames 2000; Attwell and Laughlin 2001; Lennie 2003; for review see Raichle and Mintun 2006) indicate that 60–80% of overall brain energy consumption is devoted to glutamate cycling and, hence, neural signaling processes. Such estimates leave for future consideration the demands placed on the brain's energy budget by the activity of inhibitory interneurons (Ackermann et al. 1984; McCasland and Hibbard 1997; Waldvogel et al. 2000; Chatton et al. 2003; Patel et al. 2005; Buzsáki et al. 2007) and astrocytes (Pellerin and Magistretti 1997; Magistretti and Chatton 2005). That evidence notwithstanding, it is likely to remain the case that the majority of brain energy consumption is devoted to functionally significant intrinsic activity.

1.2 Sensory Information

Complementary information on the importance of intrinsic activity comes from a consideration of sensory information. It may surprise some to learn that visual information is significantly degraded as it passes from the eye to the visual cortex. Thus, of the unlimited information available from the environment, only about 10^{10} bits/s are deposited in the retina. Because of a limited number of axons in the optic nerves (approximately 1 million axons in each), only ~6×10^6 bits/s leave the retina and only 10^4 make it to layer IV of V1 (Nørretranders 1998; Anderson et al. 2005). These data clearly leave the impression that the visual cortex receives an impoverished representation of the world, a subject of more than passing interest to those interested in the processing of visual information (Olshausen and Field 2005). In the context of this symposium, it is important to note that estimates of the bandwidth of conscious awareness itself (i.e., what we 'see') are in the range of 100 bits/sec or less (Nørretranders 1998; Anderson et al. 2005).

Reinforcing this impression of the brain's 'isolation' is the fact that the number of synapses in the lateral geniculate nucleus of the thalamus and in layer IV of primary visual cortex devoted to incoming visual information is less than 10% of the total number of synapses in both locations (Sillito and Jones 2002). Various proposals have been made concerning the interpretation of these anatomical data (Bruno and Sakmann 2006; Douglas and Martin 2007) but the fact remains that the brain must interpret, respond to and even predict environmental demands from seemingly impoverished data. An explanation for its success in doing so must lie, in significant measure, with intrinsic brain processes that link representations residing *broadly* within brain systems to incoming sensory information (Fiser et al. 2004). The challenge, of course, is how to study these intrinsic brain processes at the appropriate spatial and temporal scales.

2 Studying Intrinsic Activity

Since the introduction of electroencephalography (EEG) in humans by Hans Berger in 1929 (Berger 1929; for an English translation of this important work see Gloor 1969, it has been clear that ongoing spontaneous electrical activity is a prominent feature. In referring to the spontaneous activity in the EEG, Berger rhetorically asked (Berger 1929), "Is it possible to demonstrate the influence of intellectual work upon the human electroencephalogram, insofar as it has been reported here?"

He then concluded that: "Of course, one should not at first entertain too high hopes with regard to this, because mental work, as I explained elsewhere, adds only a small increment to the cortical work which is going on continuously and not only in the waking state." As has been demonstrated in subsequent research, extensive averaging of the EEG has been used to significantly attenuate if not eliminate this seemingly random, ongoing activity, leaving only predictably occurring, task-induced changes, or event-related potentials (ERPs) as they are known generally. This strategy is analogous to image subtraction and averaging (see Fig. 1 in Raichle 2010) with similar, unfortunate consequences for ongoing intrinsic activity.

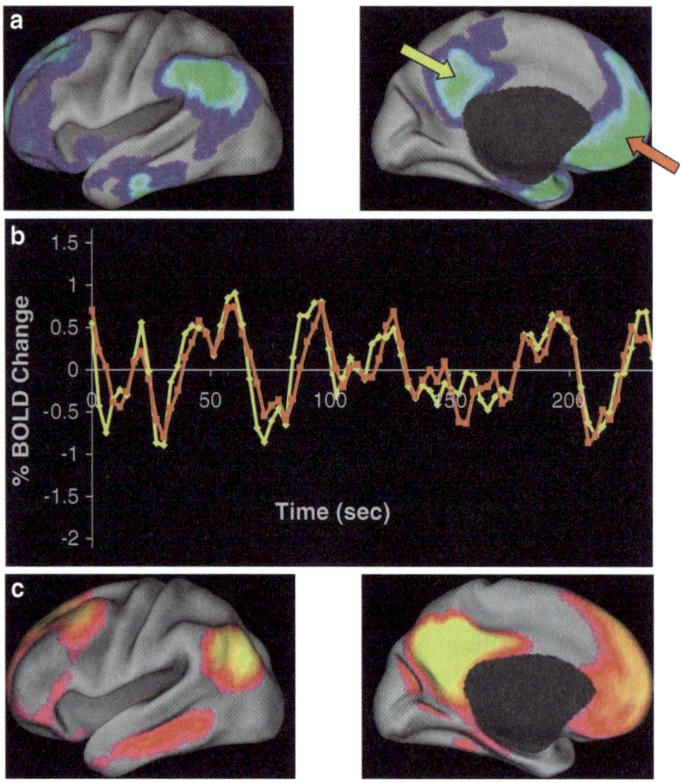

Fig. 1 Performance of a wide variety of tasks has called attention to a group of brain areas (**a**) that decrease their activity during task performance (data adapted from Shulman et al. 1997). These areas are often referred to as the brain's default mode network (DMN) after our initial work on them (Raichle et al. 2001). If one records the spontaneous fMRI BOLD signal activity in these areas in the resting state (*arrows*, **a**), what emerges is a remarkable similarity in the behavior of the signals between areas (**b**), a phenomenon originally described by Biswal and colleagues in the somatomotor cortex (1995) and later in the default mode network by Greicius and colleagues (2003). Using these fluctuations to analyze the network as a whole reveals a level of functional organization (**c**) that parallels that seen in the task-related activity decreases (**a**). (This figure was reproduced from Raichle 2010 with permission)

It was a chance observation in neuroimaging, first with PET and later with fMRI, that actually provided a new perspective on what to look for in studying the brain's intrinsic activity: the occurrence of activity decreases during the performance of goal-directed tasks compared to the resting state.

2.1 Activity Decreases from a Resting State

The first formal characterization of task-induced activity decreases from a resting state was a large meta-analysis of published PET data from our group (Shulman et al. 1997). This study generated a set of iconic images of a constellation of brain regions now generally referred to as the default mode network, or DMN (Fig. 1a), after our later paper on a *default mode of brain function* (Raichle et al. 2001). The unique identity of this group of brain regions was amply confirmed in later meta-analyses by Jeffery Binder and colleagues (1999) and Bernard Mazoyer and colleagues (2001). Similar observations are now an everyday occurrence in numerous laboratories, leaving little doubt that a specific set of brain areas decreases their activity across a remarkably wide array of task conditions when compared to a passive control condition such as visual fixation.

The finding of a network of brain areas frequently seen to decrease their activity from a resting state during goal-directed tasks (Fig. 1a) was both surprising and challenging: surprising, because the areas involved had not previously been recognized as a system in the same way we might think of the motor or visual system, and challenging, because initially it was unclear how to characterize their activity arising as it did in a passive or resting condition. Many argued, informally, that these activity decreases were simply activations present in a poorly constrained resting state.

It was clear that we needed a way to determine whether or not task-induced activity decreases were simply 'activations' in the absence of an externally directed task. To initiate our inquiry, we employed quantitative PET measurements of regional brain blood flow and oxygen consumption to define a physiologic baseline. The details of this work have been recounted on several occasions (Gusnard and Raichle 2001; Raichle et al. 2001; Raichle and Mintun 2006; Raichle and Snyder 2007) and, thus, will not be repeated here. Suffice to say that this work allowed us to move forward on the assumption that activity within the DMN did not represent conventional activations in the resting state but, rather, a new view of the organization of the brain's intrinsic activity, which we dubbed 'a default mode of brain function' (Raichle et al. 2001). It is important to note that the DMN is not unique in exhibiting both high levels of baseline metabolic activity and organized functional activity in the resting state. It is a property of all brain systems and their subcortical connections, as I will detail moving forward.

The discovery of the DMN made apparent the need for additional ways to study the large-scale intrinsic organization of the brain. A major step forward was the discovery that this large-scale network organization, including but not limited to the

DMN, could be revealed by the study of patterns of spatial coherence in the spontaneous fluctuations (i.e., noise) in the fMRI BOLD signal.

2.2 Spontaneous Fluctuations in the fMRI BOLD Signal

A prominent feature of fMRI is the noise in the raw BOLD signal, which has prompted researchers to average their data to increase signal and reduce noise. As first shown by Bharat Biswal and colleagues in the human somatomotor system (1997), a considerable fraction of this noise exhibits striking patterns of coherence within known brain systems.

The significance of this observation was brought forcefully to our attention when Michael Greicius and colleagues looked at the patterns of coherence in the DMN (2003) that were elicited by placing a region of interest in either the posterior cingulate cortex (yellow arrow, Fig. 1a) or the ventral medial prefrontal cortex (orange arrow, Fig. 1a). The resulting time-activity curves (Fig. 1b) reflected a pattern of coherence within the entire DMN (Fig. 1c). Similar patterns of resting state coherence have now been documented in most cortical systems in the human brain (for a recent review, see Fox and Raichle 2007; Smith et al. 2009) as well as their subcortical connections (Zhang et al. 2008).

A number of additional observations made about these surprising patterns of spatial coherence are of interest. First, they appear to transcend levels of consciousness, being present under anesthesia in humans (Greicius et al. 2008), monkeys (Vincent et al. 2007) and rats (Lu et al. 2007) and also during the early stages of sleep in humans (Fukunaga et al. 2006; Larson-Prior et al. 2009). These observations make it unlikely that the patterns of coherence and the intrinsic activity they represent are primarily the result of unconstrained, conscious cognition (i.e., mind-wandering or day dreaming; Christoff et al. 2009).

Second, while resting state patterns of coherence do respect patterns of anatomical connectivity in both the monkey (Vincent et al. 2007) and human brain (Zhang et al. 2008), it is clear that they are not constrained by these anatomical connections. Thus, the absence of monosynaptic connections between brain areas (e.g., right and left primary visual cortex; Vincent et al. 2007) does not preclude the existence of functional connectivity, as expressed in the maps of resting state coherence.

Third, the strength of coherence between nodes within systems varies with age and disease. Developmental changes have been particularly well demonstrated in the DMN (Fair et al. 2008). Such observations are consistent with the role of experience and, possibly, spontaneous activity itself in sculpting and maintaining these functional relationships in the human brain (Yuste 1997; Huberman et al. 2008). At the other end of the life spectrum, data suggest that the young adult pattern in the DMN may recede as one passes into the sixth decade of life and beyond (Andrews-Hanna et al. 2007), even in healthy older persons. Even more interesting are three recent studies demonstrating disruption in DMN coherence in cognitively normal older persons harboring DMN amyloid plaques (Hedden et al. 2009; Sheline et al. 2010; Sperling et al. 2009). In this regard, it should be recalled that the DMN appears to be the target of Alzheimer's disease (Buckner et al. 2005). Disruption in the resting state coherence between nodes of a system may well prove to be a very sensitive early indicator of disease (Zhang and Raichle 2010).

Finally, spontaneous fluctuations in the BOLD signal contribute significantly to variability in both evoked signals (Fox et al. 2006) and the associated behavior (Fox et al. 2007). These observations become important as we consider the neurophysiologic correlates of the spontaneous BOLD fluctuations. What follows is a brief description of the direction the neurophysiologic work is taking, beginning with the electrical correlates of the fMRI BOLD signal.

2.3 *Neurophysiology*

There has been an active effort to ascertain the electrical correlates of the fMRI BOLD signal (for summaries of this work from different perspectives, see Raichle and Mintun 2006; Khader et al. 2008; Logothetis 2008). The conclusion from this work is that the fMRI BOLD signal is best correlated with local field potentials (LFPs), that is, the complex signals arising from the integrated electrical activity in pre- and postsynaptic terminals of the brain, which are recorded with microelectrodes placed within brain tissue. Brain electrical activity recorded from the scalp with EEG or from surface of the brain with electrocorticography (ECoG) constitutes a summation of a population of LFPs. LFPs are conventionally described in terms of their band-limited frequency components (delta, 1–4 Hz; theta, 4–8 Hz; alpha, 8–12 Hz; beta, 12–24 Hz; and, gamma, >24 Hz).

Given the relationship between LFPs and BOLD, it is important to focus on those LFP phenomena that exhibit frequencies similar to that of spontaneous BOLD fluctuations (i.e., 0.01–4.0 Hz). Two LFP phenomena fall into this category: fluctuations in the power of higher frequencies (i.e., their power spectral density), where particular attention has been paid to the gamma frequency band (Leopold et al. 2003) because of its association with cognition (Fries 2009; Uhlhaas et al. 2009), and raw frequencies that approximate that of the spontaneous BOLD signal. These include the delta band (1–4 Hz), up-an-down states (~0.8 Hz; Steriade et al. 1993; Petersen et al. 2003; Hahn et al. 2006; Watson et al. 2008), and infra-slow fluctuations, or ISFs (0.01 and 0.1 Hz; Vanhatalo et al. 2004; Monto et al. 2008). ISFs are sometimes called direct current (DC) potentials, referring to their amplification requirements. ISFs are much less often recorded because of the amplifier requirements and concerns about artifacts (Khader et al. 2008). Often all three (delta, up-and-down states and ISFs) are subsumed under the designation slow cortical potentials, or SCPs (Rockstroh et al. 1989; He et al. 2008). In the present discussion, I will use the term SCPs, recognizing that it likely includes all three phenomena to an as yet unspecified extent.

The research shows that the spontaneous fluctuations in the BOLD signal are best correlated with LFP activity in the range of the SCPs (Lu et al. 2007; He et al. 2008). As is the case with the BOLD signal (see above), the spatial patterns of coherence exhibited by SCPs are maintained across levels of consciousness ranging from wakefulness to REM and slow wave sleep (He et al. 2008) and during anesthesia (Lu et al. 2007). In contrast, power in the gamma frequency band is only correlated spatially with the BOLD signal during wake and REM sleep (He et al. 2008; see also Nir et al. 2008). This finding is consistent with the role of gamma-band coherence in the mental activities associated with conscious awareness (Fries 2009; Uhlhaas et al. 2009).

Knowing that SCPs and spontaneous fluctuations in the BOLD signal are related provides a bridge to a highly relevant neurophysiologic literature on low frequency oscillations (e.g., see Rockstroh et al. 1989; Varela et al. 2001; Buzsáki and Draguhn 2004; Monto et al. 2008; Schroeder and Lakatos 2008). Emerging from this literature are several informative themes.

SCPs and quite possibly the spontaneous fluctuations in brain oxygenation as seen by fMRI BOLD and optical imaging techniques (White et al. 2009) represent fluctuations in cortical excitability (for review see Schroeder and Lakatos 2008) These fluctuations in cortical excitability have a remarkable effect on other elements of the LFP frequency spectrum, both during task-evoked activity (Monto et al. 2008) and in the resting state (He et al. 2010). SCPs also influence the spiking activity of neurons (Montemurro et al. 2008; Petermann et al. 2009). This coupling or nesting with SCPs serving an important coordinating role provides a logical structure for the integration of functional activity. Not surprisingly, the phase of the SCPs affects both evoked responses (Bishop 1933; Arieli et al. 1996; Fiser et al. 2004; Fox et al. 2006) and behavioral performance (Fox et al. 2007; Lakatos et al. 2008; Monto et al. 2008). One among many results of this functional organization might well be the emergence of conscious awareness (He and Raichle 2009). But the story does not end there.

While each major brain system can be identified by a unique pattern of spatial coherence the brain does not operate as an assemblage of independent systems. As it turns out, coordination among brain systems is aided by the fact that brain systems are not created equal. Rather, there appears to be a federation of hierarchically organized *hubs* in the brain whose influence extends across the boundaries of systems (Hagmann et al. 2008; Buckner et al. 2009). Remarkably, at the top of this hierarchy is the DMN. Thus, not surprisingly the brain is not a free-for-all among independent systems but a federation of interdependent components hierarchically organized.

SCPs provide a window on how the brain matches its predictions to changing environmental contingencies. Schroeder and Lakatos (2008) view this as one mode of attending, in which the phase of the SCPs is shifted to match the predictable patterns of incoming information, a process dubbed *phase resetting*. As a result, responses are enhanced and performance is improved (see also Bressler et al. 2008; Monto et al. 2008; Sylvester et al. 2009). This mode, and it may well be the dominant mode, occurs in a seemingly effortless manner fitting, in a sense, the idea of a default mode of brain function involving the ongoing coordinated activity of all of its systems. It provides a means of connecting the concept of an intrinsic mode of brain function designed to organize information for interpreting, responding to and even predicting environmental events (Ingvar 1985; Bar 2007) into register with the naturally occurring but ever changing regularities unfolding in the environment.

Obviously, the world in which we live is not entirely predictable, which calls for a complementary strategy in which SCPs are temporarily suppressed in a setting in which task performance requires considerable effort (i.e., goal-directed attention) because of novelty and uncertainty (for review, see Schroeder and Lakatos 2008). This is precisely the circumstance in which we have come to expect activity decreases in the DMN (Fig. 1a) and increases in brain areas associated with goal-directed attention (Corbetta and Shulman 2002; Dosenbach et al. 2008). It is of interest that this give-and-take relationship between the DMN and these latter systems is also seen in their spontaneous activity in the resting state (Fox et al. 2005, 2009; Popa et al. 2009), suggesting that the brain is continuously trying to strike a balance between the expected and the unpredictable.

2.4 Cell Biology

Pursuing an understanding of the brain's intrinsic activities need not stop with neurophysiology. Understanding the underlying cell biology is also relevant to understanding not only brain imaging signals but also what these signals are actually telling about brain function. It is beyond the scope of this brief review and the theme of this conference to cover these issues here. (Interested readers will find a review of this subject in Raichle 2010).

3 The future

As we move forward, the scope of the inquiry into the brain's intrinsic activities is poised to expand. While it may be too strong to suggest that we are facing a paradigm shift (Kuhn 1996), certainly a reorientation is taking place in terms of how we understand brain function. While this reorientation has been stimulated by brain imaging work with PET and fMRI, it has its roots in more than a century of discussions on the nature of brain functions. Integrating across the necessary levels of analysis will obviously be challenging and will demand the willingness to accept the multidisciplinary nature of the task. Help will come also from theoretical modeling approaches, where creative work has already begun (Holcman and Tsodyks 2006; Ghosh et al. 2008; Bullmore and Sporns 2009). But, just as we will not understand evoked activity without first understanding intrinsic activity, so we will not understand consciousness without first understanding non-conscious activity, for in both instances the latter dominates the former.

References

Ackermann RF, Finch DM, Babb TL, Engel J Jr (1984) Increased glucose metabolism during long-duration recurrent inhibition of hippocampal pyramidal cells. J Neurosci 4:251–264
Ames A 3rd (2000) CNS energy metabolism as related to function. Brain Res Brain Res Rev 34:42–68
Anderson CH, Van Essen DC, Olhausen BA (2005) Directed visual attention and the dynamic control of information flow. In: Itti L, Rees G, Tsotsos J (eds) Neurobiology of attention. Elsevier, San Diego, CA, pp 11–17
Andrews-Hanna JR, Snyder AZ, Vincent JL, Lustig C, Head D, Raichle ME, Buckner RL (2007) Disruption of large-scale brain systems in advanced aging. Neuron 56:924–935
Arieli A, Sterkin A, Grinvald A, Aertsen A (1996) Dynamics of ongoing activity: explanation of the large variability in evoked cortical responses. Science 273:1868–1871
Attwell D, Laughlin SB (2001) An energy budget for signaling in the grey matter of the brain. J Cereb Blood Flow Metab 21:1133–1145
Bar M (2007) The proactive brain: using analogies and associations to generate predictions. Trends Cogn Sci 11:280–289
Berger H (1929) Uber des Elektrenkephalogramm des Menschen. Arch Psychiatr Nervenkr 87:527–580
Binder JR, Frost JA, Hammeke TA, Bellgowan PS, Rao SM, Cox RW (1999) Conceptual processing during the conscious resting state. A functional MRI study. J Cogn Neurosci 11:80–95
Bishop G (1933) Cyclic changes in excitability of the optic pathway of the rabbit. Am J Physiol 103:213–224
Biswal B, Yetkin FZ, Haughton VM, Hyde JS (1995) Functional connectivity in the motor cortex of resting human brain using echo-planar MRI. Magn Reson Med 34:537–541
Biswal B, Hudetz AG, Yetkin FZ, Haughton VM, Hyde JS (1997) Hypercapnia reversibly suppresses low-frequency fluctuations in the human motor cortex during rest using echo-planar MRI. J Cereb Blood Flow Metab 17:301–308
Bressler SL, Tang W, Sylvester CM, Shulman GL, Corbetta M (2008) Top-down control of human visual cortex by frontal and parietal cortex in anticipatory visual spatial attention. J Neurosci 28:10056–10061
Brown TG (1914) On the nature of the fundamental activity of the nervous centres; together with an analysis of the conditioning of rhythmic activity in progression, and a theory of the evolution of function in the nervous system. J Physiol 48:18–46
Bruno RM, Sakmann B (2006) Cortex is driven by weak but synchronously active thalamocortical synapses. Science 312:1622–1627
Buckner RL, Snyder AZ, Shannon BJ, LaRossa G, Sachs R, Fotenos AF, Sheline YI, Klunk WE, Mathis CA, Morris JC, Mintun MA (2005) Molecular, structural, and functional characterization of Alzheimer's disease: evidence for a relationship between default activity, amyloid, and memory. J Neurosci 25:7709–7717
Buckner RL, Sepulcre J, Talukdar T, Krienen FM, Liu H, Hedden T, Andrews-Hanna JR, Sperling RA, Johnson KA (2009) Cortical hubs revealed by intrinsic functional connectivity: mapping, assessment of stability, and relation to Alzheimer's disease. J Neurosci 29:1860–1873
Bullmore E, Sporns O (2009) Complex brain networks: graph theoretical analysis of structural and functional systems. Nat Rev Neurosci 10:186–198
Buzsáki G, Draguhn A (2004) Neuronal oscillations in cortical networks. Science 304:1926–1929
Buzsáki G, Kaila K, Raichle M (2007) Inhibition and brain work. Neuron 56:771–783
Chatton JY, Pellerin L, Magistretti PJ (2003) GABA uptake into astrocytes is not associated with significant metabolic cost: implications for brain imaging of inhibitory transmission. Proc Natl Acad Sci USA 100:12456–12461

Christoff K, Gordon AM, Smallwood J, Smith R, Schooler JW (2009) Experience sampling during fMRI reveals default network and executive system contributions to mind wandering. Proc Natl Acad Sci USA 106:8719–8724
Clarke DD, Sokoloff L (1999) Circulation and energy metabolism of the brain. In: Agranoff BW, Siegel GJ (eds) Basic Neurochemistry. Molecular, cellular and medical aspects. Lippincott-Raven, Philadelphia, PA, pp 637–670
Corbetta M, Shulman GL (2002) Control of goal-directed and stimulus-driven attention in the brain. Nat Rev Neurosci 3:201–215
Dosenbach NU, Fair DA, Cohen AL, Schlaggar BL, Petersen SE (2008) A dual-networks architecture of top-down control. Trends Cogn Sci 12:99–105
Douglas RJ, Martin KA (2007) Recurrent neuronal circuits in the neocortex. Curr Biol 17: R496–500
Fair DA, Cohen AL, Dosenbach NU, Church JA, Miezin FM, Barch DM, Raichle ME, Petersen SE, Schlaggar BL (2008) The maturing architecture of the brain's default network. Proc Natl Acad Sci USA 105:4028–4032
Fiser J, Chiu C, Weliky M (2004) Small modulation of ongoing cortical dynamics by sensory input during natural vision. Nature 431:573–578
Fox MD, Raichle M (2007) Spontaneous fluctuations in brain activity observed with functional magnetic resonance imaging. Nat Rev Neurosci 8:700–711
Fox MD, Snyder AZ, Vincent JL, Corbetta M, Van Essen DC, Raichle ME (2005) The human brain is intrinsically organized into dynamic, anticorrelated functional networks. Proc Natl Acad Sci USA 102:9673–9678
Fox MD, Snyder AZ, Zacks JM, Raichle ME (2006) Coherent spontaneous activity accounts for trial-to-trial variability in human evoked brain responses. Nat Neurosci 9:23–25
Fox MD, Snyder AZ, Vincent JL, Raichle ME (2007) Intrinsic fluctuations within cortical systems account for intertrial variability in human behavior. Neuron 56:171–184
Fox MD, Zhang D, Snyder AZ, Raichle ME (2009) The global signal and observed anticorrelated resting state brain networks. J Neurophysiol 101:3270–3283
Fries P (2009) Neuronal gamma-band synchronization as a fundamental process in cortical computation. Annu Rev Neurosci 32:209–224
Fukunaga M, Horovitz SG, van Gelderen P, de Zwart JA, Jansma JM, Ikonomidou VN, Chu R, Deckers RH, Leopold DA, Duyn JH (2006) Large-amplitude, spatially correlated fluctuations in BOLD fMRI signals during extended rest and early sleep stages. Magn Reson Imaging 24:979–992
Ghosh A, Rho Y, McIntosh AR, Kötter R, Jirsa VK (2008) Noise during rest enables the exploration of the brain's dynamic repertoire. PLoS Comput Biol 4:e1000196
Gloor P (1969) Hans Berger on the electroencephalogram of man. The fourteen original reports on the human electroencephalogram. Electroencephalogr Clin Neurophysiol Suppl 28:1–350
Greicius MD, Krasnow B, Reiss AL, Menon V (2003) Functional connectivity in the resting brain: a network analysis of the default mode hypothesis. Proc Natl Acad Sci USA 100:253–258
Greicius MD, Kiviniemi V, Tervonen O, Vainionpää V, Alahuhta S, Reiss AL, Menon V (2008) Persistent default-mode network connectivity during light sedation. Hum Brain Mapp 29:839–847
Gusnard DA, Raichle ME (2001) Searching for a baseline: functional imaging and the resting human brain. Nat Rev Neurosci 2:685–694
Hagmann P, Cammoun L, Gigandet X, Meuli R, Honey CJ, Wedeen VJ, Sporns O (2008) Mapping the structural core of human cerebral cortex. PLoS Biol 6:e159
Hahn TT, Sakmann B, Mehta MR (2006) Phase-locking of hippocampal interneurons' membrane potential to neocortical up-down states. Nat Neurosci 9:1359–1361
He BJ, Raichle ME (2009) The fMRI signal, slow cortical potential and consciousness. Trends Cogn Sci 13:302–309

He BJ, Snyder AZ, Zempel JM, Smyth MD, Raichle ME (2008) Electrophysiological correlates of the brain's intrinsic large-scale functional architecture. Proc Natl Acad Sci USA 105:16039–16044

He BJ, Zempel ZM, Snyder AZ, Raichle ME (2010) The temporal structures and functional significance of scale-free brain activity. Neuron 66:353–369

Hedden T, Van Dijk KR, Sperling RA, Johnson KA, Buckner RL (2009) Disruption of functional connectivity in clinically normal older adults harboring amyloid burden. J Neurosci 29:12686–12694

Holcman D, Tsodyks M (2006) The emergence of up and down states in cortical networks. PLoS Comput Biol 2:e23

Huberman AD, Feller MB, Chapman B (2008) Mechanisms underlying development of visual maps and receptive fields. Annu Rev Neurosci 31:479–509

Ingvar DH (1985) Memory of the future: an essay on the temporal organization of conscious awareness. Hum Neurobiol 4:127–136

James W (1890) The principles of psychology. Henry Holt & Company, New York

Khader P, Schicke T, Röder B, Rösler F (2008) On the relationship between slow cortical potentials and BOLD signal changes in humans. Intl J Psychophysiol 67:252–261

Kuhn TS (1996) Structure of scientific revolutions. University of Chicago Press, Chicago, IL

Lakatos P, Karmos G, Mehta AD, Ulbert I, Schroeder CE (2008) Entrainment of neuronal oscillations as a mechanism of attentional selection. Science 320:110–113

Larson-Prior LJ, Zempel JM, Nolan TS, Prior FW, Snyder AZ, Raichle ME (2009) Cortical network functional connectivity in the descent to sleep. Proc Natl Acad Sci USA 106:4489–4494

Lennie P (2003) The cost of cortical computation. Curr Biol 13:493–497

Leopold DA, Murayama Y, Logothetis NK (2003) Very slow activity fluctuations in monkey visual cortex: implications for functional brain imaging. Cereb Cortex 13:422–433

Llinas R (2001) I of the vortex. From neurons to self. MIT, Cambridge, MA

Logothetis NK (2008) What we can do and what we cannot do with fMRI. Nature 453:869–878

Lu H, Zuo Y, Gu H, Waltz JA, Zhan W, Scholl CA, Rea W, Yang Y, Stein EA (2007) Synchronized delta oscillations correlate with the resting-state function MRI signal. Proc Natl Acad Sci USA 104:18265–18269

Magistretti PJ, Chatton JY (2005) Relationship between L-glutamate-regulated intracellular Na+ dynamics and ATP hydrolysis in astrocytes. J Neural Transm 112:77–85

Mazoyer B, Zago L, Mellet E, Bricogne S, Etard O, Houdé O, Crivello F, Joliot M, Petit L, Tzourio-Mazoyer N (2001) Cortical networks for working memory and executive functions sustain the conscious resting state in man. Brain Res Bull 54:287–298

McCasland JS, Hibbard LS (1997) GABAergic neurons in barrel cortex show strong, whisker-dependent metabolic activation during normal behavior. J Neurosci 17:5509–5527

Montemurro MA, Rasch MJ, Murayama Y, Logothetis NK, Panzeri S (2008) Phase-of-firing coding of natural visual stimuli in primary visual cortex. Curr Biol 18:375–380

Monto S, Palva S, Voipio J, Palva JM (2008) Very slow EEG fluctuations predict the dynamics of stimulus detection and oscillation amplitudes in humans. J Neurosci 28:8268–8272

Nir Y, Mukamel R, Dinstein I, Privman E, Harel M, Fisch L, Gelbard-Sagiv H, Kipervasser S, Andelman F, Neufeld MY, Kramer U, Arieli A, Fried I, Malach R (2008) Interhemispheric correlations of slow spontaneous neuronal fluctuations revealed in human sensory cortex. Nat Neurosci 11:1100–1108

Nørretranders T (1998) The user illusion: cutting consciousness down to size. Viking Press, New York NY

Olshausen BA, Field DJ (2005) How close are we to understanding v1? Neural Comput 17:1665–1699

Patel AB, de Graaf RA, Mason GF, Rothman DL, Shulman RG, Behar KL (2005) The contribution of GABA to glutamate/glutamine cycling and energy metabolism in the rat cortex in vivo. Proc Natl Acad Sci USA 102:5588–5593

Pellerin L, Magistretti PJ (1997) Glutamate uptake stimulates Na+, K+−ATPase activity in astrocytes via activation of a distinct subunit highly sensitive to ouabain. J Neurochem 69:2132–2137
Petermann T, Thiagarajan TC, Lebedev MA, Nicolelis MA, Chialvo DR, Plenz D (2009) Spontaneous cortical activity in awake monkeys composed of neuronal avalanches. Proc Natl Acad Sci USA 106:15921–15926
Petersen CC, Hahn TT, Mehta M, Grinvald A, Sakmann B (2003) Interaction of sensory responses with spontaneous depolarization in layer 2/3 barrel cortex. Proc Natl Acad Sci USA 100:13638–13643
Popa D, Popescu AT, Paré D (2009) Contrasting activity profile of two distributed cortical networks as a function of attentional demands. J Neurosci 29:1191–1201
Raichle ME (2010) Two views of brain function. Trends Cogn Sci 14:180–190
Raichle ME, Mintun MA (2006) Brain work and brain imaging. Annu Rev Neurosci 29:449–476
Raichle M, Snyder AZ (2007) A default mode of brain function: a brief history of an evolving idea. Neuroimage 37:1083–1090
Raichle ME, MacLeod AM, Snyder AZ, Powers WJ, Gusnard DA, Shulman GL (2001) A default mode of brain function. Proc Natl Acad Sci USA 98:676–682
Rockstroh B, Elbert T, Canavan AG, Lutzenberger W, Birbaumer N (1989) Slow cortical potentials and behavior. Urban & Schwarzenberg, Baltimore, MD
Schroeder CE, Lakatos P (2008) Low-frequency neuronal oscillations as instruments of sensory selection. Trends Neurosci 32:9–18
Sheline YI, Raichle ME, Snyder AZ, Morris JC, Head D, Wang S, Mintun MA (2010) Amyloid plaques disrupt resting state default mode network connectivity in cognitively normal elderly. Biol Psychiatry 67:584–587
Sherrington CS (1906) The integrative action of the nervous system. Yale University Press, New Haven, CT
Shulman GL, Fiez JA, Corbetta M, Buckner RL, Miezin FM, Raichle ME, Petersen SE (1997) Common blood flow changes across visual tasks: II Decreases in cerebral cortex. J Cogn Neurosci 9:648–663
Sibson NR, Dhankhar A, Mason GF, Behar KL, Rothman DL, Shulman RG (1997) In vivo 13C NMR measurements of cerebral glutamine synthesis as evidence for glutamate-glutamine cycling. Proc Natl Acad Sci USA 94:2699–2704
Sibson NR, Dhankhar A, Mason GF, Rothman DL, Behar KL, Shulman RG (1998) Stoichiometric coupling of brain glucose metabolism and glutamatergic neuronal activity. Proc Natl Acad Sci USA 95:316–321
Sillito AM, Jones HE (2002) Corticothalamic interactions in the transfer of visual information. Philos Trans R Soc Lond B Biol Sci 357:1739–1752
Smith SM, Fox PT, Miller KL, Glahn DC, Fox PM, Mackay CE, Filippini N, Watkins KE, Toro R, Laird AR, Beckmann CF (2009) Correspondence of the brain's functional architecture during activation and rest. Proc Natl Acad Sci USA 106:13040–13045
Sokoloff L, Mangold R, Wechsler R, Kennedy C, Kety SS (1955) The effect of mental arithmetic on cerebral circulation and metabolism. J Clin Invest 34:1101–1108
Sperling RA, Laviolette PS, O'Keefe K, O'Brien J, Rentz DM, Pihlajamaki M, Marshall G, Hyman BT, Selkoe DJ, Hedden T, Buckner RL, Becker JA, Johnson KA (2009) Amyloid deposition is associated with impaired default network function in older persons without dementia. Neuron 63:178–188
Steriade M, Nunez A, Amzica F (1993) A novel slow ($<$ 1 Hz) oscillation of neocortical neurons in vivo: depolarizing and hyperpolarizing components. J Neurosci 13:3252–3265
Sylvester CM, Shulman GL, Jack AI, Corbetta M (2009) Anticipatory and stimulus-evoked blood oxygenation level-dependent modulations related to spatial attention reflect a common additive signal. J Neurosci 29:10671–10682
Uhlhaas PJ, Pipa G, Lima B, Melloni L, Neuenschwander S, Nikolić D, Singer W (2009) Neural synchrony in cortical networks: history, concept and current status. Front Integr Neurosci 3:17

Vanhatalo S, Palva JM, Holmes MD, Miller JW, Voipio J, Kaila K (2004) Infraslow oscillations modulate excitability and interictal epileptic activity in the human cortex during sleep. Proc Natl Acad Sci USA 101:5053–5057
Varela F, Lachaux JP, Rodriguez E, Martinerie J (2001) The brainweb: phase synchronization and large-scale integration. Nat Rev Neurosci 2:229–239
Vincent JL, Patel GH, Fox MD, Snyder AZ, Baker JT, Van Essen DC, Zempel JM, Snyder LH, Corbetta M, Raichle ME (2007) Intrinsic functional architecture in the anaesthetized monkey brain. Nature 447:83–86
Waldvogel D, van Gelderen P, Muellbacher W, Ziemann U, Immisch I, Hallett M (2000) The relative metabolic demand of inhibition and excitation. Nature 406:995–998
Watson BO, MacLean JN, Yuste R (2008) UP states protect ongoing cortical activity from thalamic inputs. PLoS ONE 3:e3971
White BR, Snyder AZ, Cohen AL, Petersen SE, Raichle ME, Schlaggar BL, Culver JP (2009) Resting-state functional connectivity in the human brain revealed with diffuse optical tomography. Neuroimage 47:148–156
Yuste R (1997) Introduction: spontaneous activity in the developing central nervous system. Semin Cell Dev Biol 8:1–4
Yuste R, MacLean JN, Smith J, Lansner A (2005) The cortex as a central pattern generator. Nat Rev Neurosci 6:477–483
Zhang D, Raichle ME (2010) Disease and the brain's dark energy. Nat Rev Neurol 6:15–28
Zhang D, Snyder AZ, Fox MD, Sansbury MW, Shimony JS, Raichle ME (2008) Intrinsic functional relations between human cerebral cortex and thalamus. J Neurophysiol 100:1740–1748

Beyond Libet: Long-Term Prediction of Free Choices from Neuroimaging Signals

John-Dylan Haynes

Abstract Since the famous experiments by Benjamin Libet it has been debated whether brain signals can precede and predict the outcome of decisions even before subjects believe to be consciously making up their mind. Now it is possible to directly assess such choice-predictive neural signals using multivariate pattern classification. We have shown that the specific outcome of free choices between different plans can be read out of brain activity even several seconds before it is made. Thus, an unconscious causal chain of events can occur outside subjective awareness even before a subject makes up their mind. The nature and computational role of the choice-predictive signals still requires further investigation. Most importantly, the link to "background" brain activity is currently unclear.

1 Introduction

It is a common folk-psychological intuition that we can freely choose between different behavioral options. Even a simple, restricted movement task with only a single degree of freedom can be sufficient to yield this intuition, say in an experiment where a subject is asked to "move a finger at some point of his own choice". Although such a simple decision might not be perceived to be as important as, for example, a decision to study at one university or another, most subjects feel it is a useful example of a specific type of freedom that is often experienced when making decisions. They have the impression that the outcome of many decisions is not predetermined at the time they are felt to be made, and instead they are still "free" to choose one way or the other.

J.-D. Haynes (✉)
Bernstein Center for Computational Neuroscience Berlin, Charité – Universitätsmedizin, Berlin, Germany
and
Max-Planck-Institute for Human Cognitive and Brain Sciences, Leipzig, Germany
e-mail: haynes@bccn-berlin.de

S. Dehaene and Y. Christen (eds.), *Characterizing Consciousness: From Cognition to the Clinic?* Research and Perspectives in Neurosciences,
DOI 10.1007/978-3-642-18015-6_10,

This belief in the freedom of decisions is fundamental to our human self-concept. It is so strong that it is generally maintained even though it contradicts several other core beliefs. For example, freedom appears to be incompatible with the nature of our universe. The deterministic, causally closed physical world seems to stand in the way of "additional" and "unconstrained" influences on our behavior from mental faculties that exist beyond the laws of physics. Interestingly, in most people's (and even in some philosophers'[1]) minds, the incompatible beliefs in free will and in determinism co-exist happily without any apparent conflict. One reason why most people do not perceive this as a conflict might be that our belief in freedom is so deeply embedded in our everyday thoughts and behavior that the rather abstract belief in physical determinism is simply not strong enough to compete. The picture changes, however, with direct scientific demonstrations that our choices are determined by the brain. People are immensely fascinated by scientific experiments that directly expose how our seemingly free decisions are systematically related to prior brain activity.

In a seminal experiment, Benjamin Libet and colleagues (1983; 1985) investigated the temporal relationship between brain activity and a conscious intention to perform a simple voluntary movement. Subjects viewed a "clock" that consisted of a light point moving on a circular path rotating once every 2.56 s. Subjects were asked to flex a finger at a freely chosen point in time and to remember and report the position of the moving light point when they first felt the urge to move. The reported position of the light could then be used to determine the time when the person consciously formed their intention, a time subsequently called "W" as a shorthand for the conscious experience of "wanting" or "will." Libet recorded EEG signals from movement-related brain regions while subjects were performing this task. It had previously been known that negative deflections of the EEG signal could be observed immediately preceding voluntary movements (Kornhuber and Deecke 1965). These so-called readiness potentials (RPs) originate from a region of cortex known as the supplementary motor cortex (SMA) that is involved in motor preparation. Libet and colleagues were interested in whether the RP might begin to arise even before the person had made up his mind to move. Indeed they found that the RP began to arise a few 100 ms before the "feeling of wanting" entered awareness. This systematic temporal precedence of brain activity before a freely timed decision was subsequently taken as evidence that the brain had made the decision to move *before* this decision entered awareness. It was proposed that the RP reflects the primary cortical site where the decision to move is made (Eccles 1982).

Due to their far-reaching implications that unconscious brain processes might cause what appears to be a free choice, Libet's ground-breaking experiments immediately met with severe criticism. Following the analysis of Hume (1777), two empirical criteria are required to argue for a causal relationship between two events, for example, event B (brain) causing event W (will). First, there has to be a temporal precedence of B before W, and second there has to be a constant connection between events B and W. It has been debated whether Libet's experiments fulfill

[1]The author is an incompatibilist.

either of these criteria. Several authors have questioned whether there is indeed a temporal precedence between RP and intention, in particular by arguing that the timing judgments are unreliable (Breitmeyer 1985; Van de Grind 2002). It has long been known that there are substantial inaccuracies in determining the timing and position of moving objects (Moutoussis and Zeki 1997; Rollman 1985; Van de Grind 2002; Wundt 1904). Thus, the choice of a moving light point to report the timing is far from optimal.

A different line of arguments addresses the constant connection between B and W. Libet reported data averaged across a number of trials. Although this shows that, on average, there is a RP before the urge to move, it does not show whether this holds for every single trial, which would be necessary to provide evidence for a constant connection. For example, the early onset of the RP might be an artefact of temporal smearing and might reflect only the onset of the earliest urges to move (Trevena and Miller 2002). This possibility could only be assessed by measuring the onset time of individual RPs, which is a particularly challenging signal processing problem that requires advanced decoding algorithms (Blankertz et al. 2003).

A further important shortcoming of Libet's experiment is that it only investigates RPs, which means it is restricted to signals originating from movement-related brain regions. This experimental design leaves unclear how other areas might contribute to the build-up of decisions, which is particularly important because several other regions of prefrontal cortex have frequently been shown to be involved in free choice situations (e.g., Deiber et al. 1991), although it remains unclear to what degree they are involved in preparing a decision. Another shortcoming of RPs is that they only emerge in a narrow time window immediately preceding a movement, leaving unclear whether they indeed reflect the earliest stage where a decision is cortically prepared. In fact it has been argued that the close temporal proximity of RP and conscious awareness of the urge to move means that these two processes are scientifically indistinguishable (Merikle and Cheesman 1985).

Taken together, some of the problems with the original Libet experiment could be overcome by investigating whether other brain regions might begin to prepare a decision across longer time spans. Interestingly, even before the original Libet experiments, it had been shown that the prefrontal cortex prepares voluntary movements across longer periods than are visible from the RP alone (Groll-Knapp et al. 1977). Thus, activity in prefrontal brain regions might be a much better predictor of the outcome of decisions than RPs. However, to date, studies on voluntary movement preparation in prefrontal cortex have not simultaneously measured the timing of the self-paced urge to move along with the corresponding brain activity.

2 The Modified Libet Experiment

To overcome these and other shortcomings of the Libet experiments, we performed a novel variant of the original task (Soon et al. 2008). We used functional magnetic resonance imaging (fMRI), a technique that measures changes in the oxygenation level of blood that are in turn caused by neural activity. FMRI has a much higher spatial resolution than EEG. It uses a measurement grid with a resolution of around 3 mm to independently measure brain activity at each position in the brain. Because the fMRI signal has a low *temporal* resolution (typically around 0.5 Hz) and lags several seconds behind neural activity, it does not allow one to resolve the fine-grained cascade of neural processes in the few 100 ms just before the will enters awareness. However, it is highly suitable for looking back from the W event at each position in the brain and across longer time spans. Our focus on longer time spans and the low sampling rate of the fMRI signal enabled us to relax our requirement for temporal precision of the timing judgment, thus overcoming a severe limitation of Libet's original experiments. We replaced the rotating clock with a randomized stream of letters that updated every 500 ms. Subjects had to report the letter that was visible on the screen when they made their conscious decision. This mode of report has the additional advantage of being unpredictable, which minimizes systematic preferences for specific clock positions.

Subjects were asked to freely decide between two response buttons while lying in an MRI scanner (Fig. 1a). They fixated on the center of the screen where the stream of letters was presented. While viewing the letter stream, they were asked to relax and freely decide at some point in time to press either the left or right button. In parallel, they should remember the letter presented when their decision to move reached awareness. After subjects made up their mind and pressed their freely chosen response button, a "response mapping" screen appeared where subjects used a second button press to indicate at which time they had made their decision. This response-mapping screen showed three letters plus a hash symbol (#) arranged randomly on the four corners of an imaginary square centered on fixation. Each of these positions corresponded to one of four buttons operated by the left and right index and middle fingers. Subjects were asked to press the button corresponding to the letter that was visible on the screen when they consciously made their decision. When the letter was not among those presented on the screen, they were asked to press the button corresponding to the hash symbol. Then, after a delay, the letter stream started again and a new trial began. Note that, due to the randomization of the position of letters in the response mapping screen, the second response was uncorrelated with the first, freely chosen response. Importantly, to facilitate spontaneous behavior, we did not ask subjects to balance the left and right button selections, which would require keeping track of the distribution of button selections in memory and would also encourage preplanning of choices. Instead, we selected subjects who spontaneously chose a balanced number of left and right button presses without prior instruction based on a behavioral selection test before scanning.

3 Decoding Choices from Brain Activity Patterns

An important innovation was that we used a "decoder" to predict, from his brain activity, how a subject was going to decide (see Fig. 1b). For each time point preceding the intention, we examined whether a given brain region carried

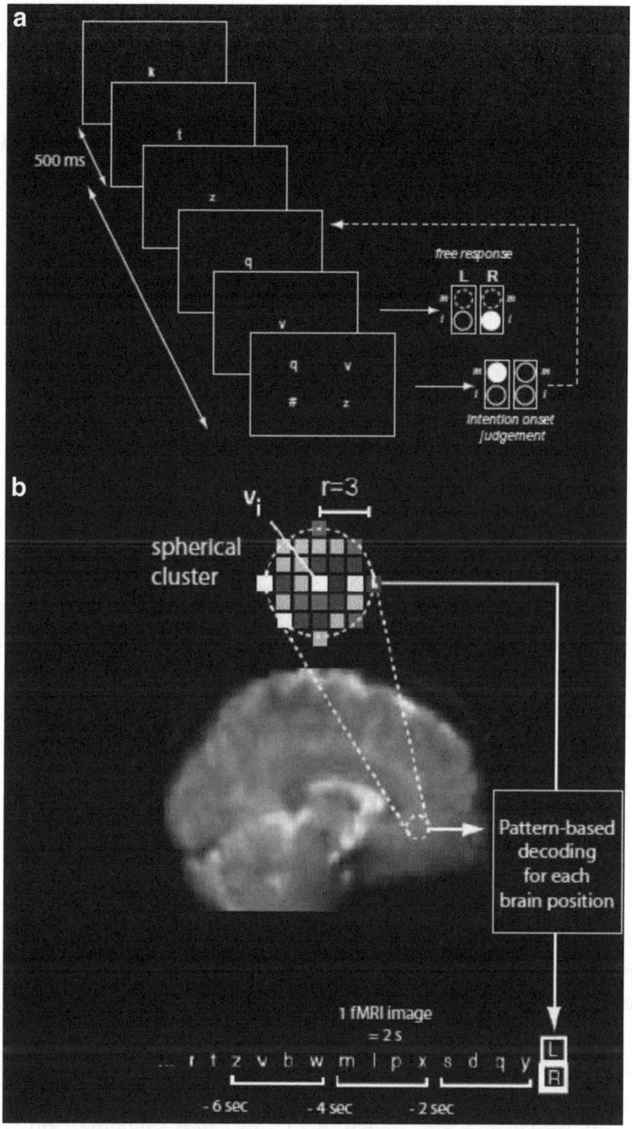

Fig. 1 (**a**) The revised Libet task. Subjects are given two response buttons, one for the left and one for the right hand. In parallel, there is a stream of letters on the screen that changes every 500 ms. They are asked to relax and to decide at some spontaneous point of their own choosing to press either the left or the right button. Once the button is pressed, they are asked to report which letter was on the screen when they made up their mind. (**b**) Pattern-based decoding and prediction of decisions ahead of time. Using a searchlight technique (Kriegeskorte et al. 2006; Haynes et al. 2007; Soon et al. 2008), we assessed for each brain region and each time point preceding the decision whether it is possible to decode the choice ahead of time. Decoding is based on small, local spherical clusters of voxels that form three-dimensional spatial patterns. It allowed us to systematically investigate which brain regions had predictive information at each time point preceding the decision

information related to the specific outcome of a decision, that is the urge to press *either* a left *or* a right button, rather than reflecting unselective motor preparation. To understand the advantage of "decoding," it can help to review the standard analysis techniques in fMRI studies. Most conventional neuroimaging analyses perform statistical analyses on one position in the brain at a time and then proceed to the next position (Friston et al. 1995). This yields a map of statistical parameters that plots how strong a certain effect is expressed at each *individual* position in the brain. However, this neglects any information that is present in the distributed spatial patterns of fMRI signals. Typically, the raw data are also spatially smoothed, so any fine-grained spatial patterning is lost. It has recently emerged, however, that these fine-grained fMRI patterns contain information that is highly predictive of the detailed contents of a person's thoughts (Kamitani and Tong 2005; Haynes and Rees 2005, 2006). This finding is in accord with a common view that each region of the brain encodes information in a distributed spatial pattern of activity (Tanaka 1997). This information is lost for conventional analyses. The full information present in brain signals can only be extracted by jointly analyzing multiple locations using pattern-based decoding algorithms. Conventional analyses can only reveal whether a brain area is more or less active during a task (say immediately preceding a decision). In contrast, we used the novel pattern-based decoding analyses not to investigate the overall level of activity but to extract the *maximal amount* of predictive information contained in the fine-grained spatial pattern of activity. This information allows one to predict the *specific* choice a subject is going to make on each trial.

To first validate our method, we investigated from which brain regions the specific decision could be decoded *after* it had been made and the subject was already executing the motor response (Fig. 2, top). This procedure served as a sanity check because it is clear that one would expect to find the decision to be encoded in the motor cortex. Thus, for each brain area and each time point after the decision, we assessed whether it was possible to decode from the spatial pattern of brain signals which motor response the subject was *currently* executing. As expected, two brain regions encoded the outcome of the subject's decision during the execution phase. These were primary motor cortex and SMA. Thus, the sanity check demonstrates the validity of the method. Please note that, as expected, the informative fMRI signals are delayed by several seconds relative to the decision due to the delay of the hemodynamic response.

Next we addressed the key question of this study, whether any brain region encoded the subject's decision *ahead* of time. We found that, indeed, two brain regions predicted prior to the conscious decision whether the subject was about to choose the left or right response, even though the subject did not know yet which way they were about to decide (Fig. 2, bottom). The first region was in frontopolar cortex (FPC), BA10. The predictive information in the fMRI signals from this brain region was already present 7 s prior to the subject's decision. This period of 7 s is a conservative estimate that does not yet take into account the delay of the fMRI response with respect to neural activity. Because this delay is several seconds, the predictive neural information will have preceded the conscious decision by up to

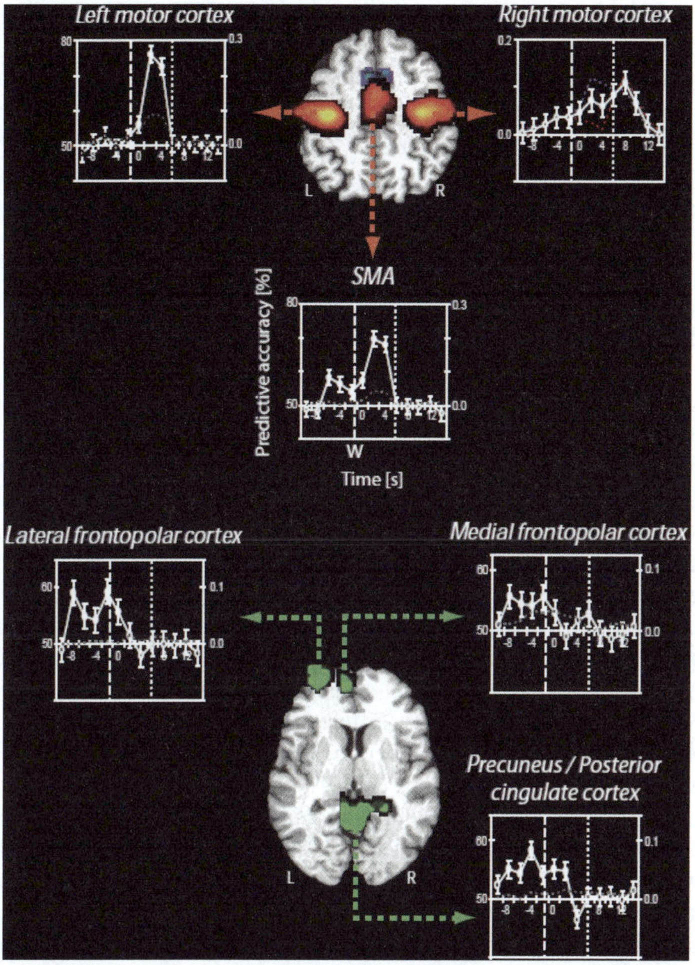

Fig. 2 (*Top*) First, we assessed which brain regions had information about a subject's decision after it had been made and the subject was currently pressing the button corresponding to his choice. As expected, this yielded information in motor cortex and supplementary motor cortex. (*Bottom*) Second, we assessed which brain regions had predictive information about a subject's decision even before the subject knew how he was going to decide. This yielded regions of the frontopolar cortex and precuneus/posterior cingulate cortex, which already had predictive information 7 s before the decision was made

10 s. There was a second predictive region located in parietal cortex (PC), stretching from the precuneus into posterior cingulate cortex. It is important to note that there is no *overall* signal increase in the frontopolar and precuneus/posterior cingulate during the preparation period. Rather, the predictive information is encoded in the spatial pattern of fMRI responses, which is presumably why it has only rarely been noticed before. Please note that, due to the temporal delay of the

hemodynamic response, the small lead times in SMA/pre-SMA of up to several 100 ms reported in previous studies (Libet et al. 1983; Haggard and Eimer 1999) are below the temporal resolution of our method. Hence, we cannot exclude that other regions contain predictive information in the short period *immediately preceding* the intention.

3.1 The Role of Brodman Area 10

The finding of unconscious, predictive brain activity patterns in Brodman area 10 (BA 10) is interesting because this area is not normally discussed in connection with free choices, presumably due to the fact that conventional analyses will only pick up regions with *overall* changes in activity but not regions where only the *patterning* of the signal changes in a choice-specific fashion. However, it has been repeatedly demonstrated using other tasks that BA 10 plays an important role encoding and storage of intentions. It has long been known that lesions to BA 10 lead to a loss of prospective memory, thus disrupting the ability to hold action plans in memory for later execution (Burgess et al. 2001). A previous study from our group showed that BA 10 also stores intentions across delay periods after they have reached consciousness, especially if there is a delay between decision and execution (Haynes et al. 2007). Although BA 10 has only rarely been implicated in preparation of voluntary actions, a direct comparison across different brain regions has revealed that the earliest cortical region exhibiting preparatory signals before voluntary movements is the frontopolar cortex (Groll-Knapp et al. 1977). BA 10 is also cytoarchitectonically very special. It has a very low cell density, but each cell forms a large number of synapses, meaning that it is a highly associative brain region (Ramnani and Owen 2004), which might allow for locally recurrent processing that could support the storage of action plans in working memory. Furthermore, BA 10 is believed to be the area that has most disproportionately grown in size in humans compared to non-human primates (Ramnani and Owen 2004).

3.2 Two Preparatory Circuits: "What" Versus "When"

When taking a closer look, it becomes apparent that multi-choice versions of the Libet experiment involve not just one but two decisions to be made (Haggard and Eimer 1999; Soon et al. 2008). On the one hand a decision needs to be made as to *when* to decide: on the other hand a decision has to be made as to *which button* to choose. Brass and Haggard (2008) have referred to these as "when" and "what" decisions. So far we have decoded the "what" decisions, so next we conducted a further decoding analysis where we assessed to what degree the timing of the decision (as opposed to its outcome) could be decoded. The time of conscious intention could be significantly predicted from pre-SMA and SMA. The earliest decodable information on timing was available 5 s before a decision, which might suggest that the brain begins to prepare self-paced decisions through two independent networks that only converge at later stages of processing. The classical Libet experiments, which were primarily concerned with "when" decisions, found short-term predictive information in the SMA. This finding is compatible with our prediction of the timing from pre-SMA and SMA. In contrast, as our results show, a "what" decision is prepared much earlier and by a much more extended network in the brain.

4 Sanity Checks

Our findings point towards long-leading brain activity that is predictive of the outcome of a decision even before the decision reaches awareness. This is a striking finding and thus it is important to critically discuss several possible sources of artefacts and alternative interpretations. Of particular interest is to make sure that the report of the timing is correct and that the information does not reflect a carry-over from previous trials.

4.1 *Early Decision: Late Action?*

One question is whether the subjects are really performing the task correctly. For example, they might decide early, say at the beginning of the trial, which button to press, and then simply wait for a few seconds to execute their response. This could be the case if say the entire group of subjects had been grossly disregarding the instructions. A similar argument has already been made against the Libet experiment. It is conceivable that, as the decision outcome *gradually* enters awareness, subjects adopt a very conservative criterion for their report and wait for the awareness to reach its "peak" intensity (Latto 1985; Ringo 1985). Fortunately, there are reasons that make it implausible that subjects simply waited to report that the decision had already begun to reach awareness. In situations where subjects know which button they are going to press, the corresponding movement is already prepared all the way up to the primary motor cortex. In contrast, in our study, the motor cortex contains information only at a very late stage of processing, following the conscious decision of which movement to make, suggesting that subjects did not decide early and then simply wait.

4.2 *Carry-Over from Previous Trial?*

Importantly, it is also possible to rule out that the early prediction presumably reflects a carryover of information from the previous trial. First, the distribution of response sequences clearly resembles an exponential distribution without sequential order, as would be expected if subjects decide randomly from trial to trial which button to press. This is presumably because, in contrast to previous studies, we did not ask subjects to balance left and right button presses across trials, thus encouraging decisions that were independent of previous trials. Also, in our experiments, subjects often took a long time until they made a decision, which might explain why subjects behaved more randomly than in traditional random choice experiments where subjects systematically violate randomness when explicitly asked to rapidly generate random sequences (Nickerson 2002). Second, our chosen statistical analysis method, fitting a so-called finite impulse response function, is designed to separate the effects of the current trial from the previous and the following trial. This approach is highly efficient as long as both types of responses are equally frequent, with variable inter-trial intervals, as here. Third, the early onset of predictive information in prefrontal and parietal regions cannot be explained by any trailing BOLD signals from the previous trials. The onset of information occurs approximately 12 s after the previous trial, which is far beyond the relaxation time of the hemodynamic response. Also, the predictive information increases with temporal distance from the previous trial, which is not compatible with the information being an overlap from the previous trial. Fourth, time points that overlap into the next trial also revealed no carry-over of information. The high predictive accuracy preceding the decision reflects prospective information encoded in prefrontal and parietal cortex related to the decision in the current trial.

5 Implications for the Free-Will Debate?

Our study shows that the brain can begin to unconsciously prepare decisions several seconds before it reaches awareness. Does our study thus have any novel implications for the debate on free will that has so far heavily relied on Libet's experiments? The potential implications of Libet's experiments for free will have been discussed at great length in the literature, which has helped define what the contribution of such simple free choice paradigms might be. Obviously they do not address real-world decisions that have high motivational importance, they are not based on long-term reward expectations and they do not involve complex reasoning. Our and Libet's decisions have only little motivational salience for the individual and are experienced as random rather than being based on in-depth trial to trial reasoning. However, our and Libet's findings do address one specific intuition regarding free will: the naïve folk-psychological intuition that, at the time when we make a decision, the outcome of this decision is free and not fully determined by brain activity. As discussed above, this intuition is scientifically implausible anyway, simply because it stands in contradiction to our belief in a deterministic universe. However, the direct demonstration that brain activity predicts the outcomes of decisions before they reach awareness has additional persuasive power. Dissociations between awareness and brain processing are nothing unusual; they have been demonstrated in motor control before (Fourneret and Jeannerod 1998). What our findings now show is that a whole cascade of unconscious brain processes unfolds across several seconds and helps prepare subjectively free, self-paced decisions.

6 Causality?

An important point that needs to be discussed is to what degree our findings support any causal relationship between brain activity and the conscious will. For the criterion of *temporal precedence,* there should be no doubt that our data finally demonstrate that brain activity can predict a decision long before it enters awareness. A different point is the criterion of *constant connection.* For a constant connection, one would require that the decision could be predicted with 100% percent accuracy from prior brain activity. Libet's original experiments were based on averages, so no statistical assessment can be made about the accuracy with which decisions can be predicted. Our prediction of decisions from brain activity is statistically reliable but far from perfect. The predictive accuracy of around 60% can be substantially improved if the decoding is custom-tailored for each subject. However, even under optimal conditions, this percentage is far from 100%. There could be several explanations. One possibility is that the inaccuracy stems from imperfections in our ability to measure neural signals. Due to the limitations of fMRI in terms of spatial and temporal resolution, it is clear that the information we can measure can only reflect a strongly impoverished version of the information available from a direct measurement of the activity in populations of neurons in the predictive areas. A further source of imperfection is that an optimal decoding approach needs a large (ideally infinite) number of training samples to learn exactly what the predictive patterns should be. In contrast, the slow sampling rate of fMRI imposes limitations on the training information available. So, even if the populations of neurons in these areas would, in principle, allow a perfect prediction, our ability to extract this information would be severely limited. However, these limitations cannot be used to argue that one day, with better methods, the prediction will be perfect; this would constitute a mere "promissory" prediction. Importantly, a different interpretation could be that the inaccuracy simply reflects the fact that the early neural processes might *in principle* simply not be fully, but only partially, predictive of the outcome of the decision. In this view, even full knowledge of the state of activity of populations of neurons in the frontopolar cortex and in the precuneus would not permit us to fully predict the decision. In that case, the signals have the form of a biasing signal that influences the decision to a degree, but additional influences at later time points might still play a role in shaping the decision. Until a perfect predictive accuracy has been reached in an experiment, both interpretations – incomplete prediction and incomplete determination – remain possible.

7 Future Perspectives

An important question for future research is whether the signals we observed are indeed decision-related. This question might sound strange given that they predict the choices. However, this early information could hypothetically also be the consequence of stochastic, fluctuating background activity in the decision network (Eccles 1985), similar to the known fluctuations of signals in early visual cortex (Arieli et al. 1996). In this view, the processes relevant for the decision would occur late, say in the last second before the decision. In the absence of any "reasons" for deciding for one or the other option, the decision network might need to break the symmetry, for example, by using stochastic background fluctuations in the network. If the fluctuations in the network are in one subspace, the decision could be pushed towards "left" and, if the fluctuations are in a different subspace, the decision could be pushed towards "right." But how could fluctuations at the time of the conscious decision be already reflected 7 s before? One possibility is that the temporal autocorrelation of the fMRI signal smears the ongoing fluctuations across time. However, the fMRI signal itself is presumably not causally involved in decision-making; it is only an indirect way of measuring the *neural* processes leading up to the decision. Thus the relevant question is the temporal autocorrelation of neural signals, which seems incompatible with a timescale of 7–10 s. Nevertheless, in future experiments we aim to investigate even further how tightly the early information is linked to the decision. One prediction of the slow background fluctuation model is that the outcome of the decision would be predictable even in cases where a subject does not know that he is going to have to make a decision or where a subject does not know what a decision is going to be about. This would point towards a predictive signal that does not directly computationally contribute to decision making.

A further interesting point for future research is the comparison of self-paced with rapid decisions that occur in response to sudden and unpredictable external events. At first sight, it seems implausible that rapid, responsive decisions could be predicted ahead of time. How would we be able to drive a car on a busy road if it always took us a minimum of 7 s to make a decision? However, even unpredictable decisions are likely to be determined by "cognitive sets" or "policies" that are likely to have a much longer half-life in the brain than a mere 7 s.

Finally, it would be interesting to investigate whether decisions can be predicted in real-time, before a person knows how he is going to decide. Such a real-time "decision prediction machine" (DP machine) would us allow to turn certain thought experiments (Marks 1985; Chiang 2005) into reality, for example, by testing whether people can guess above chance which future choices are predicted by their current brain signals even though a person might not have yet made up his mind. Such forced-choice judgments would be helpful in revealing whether there is evidence for subtle, decision-related information that might enter a person's awareness at an earlier stage than would be apparent in the conventional Libet tasks (Marks 1985). A different experiment could be to ask a person to press a button at a time point of his own choosing, with the one catch that he is not allowed to press it when a lamp lights up (Chiang 2005). Using real-time decoding techniques, it might then be possible to predict the impending decision to press the button and to control the lamp to prevent the action. The phenomenal experience of performing such an experiment would be interesting. For example, if the prediction is early enough, the subject is not even aware that he is about to make up his mind and should have the impression that the light is flickering on and off randomly. It would be possible to use the DP machine to inform the subject of the impending decision and get him to "veto" his action and not press a button. Currently, such "veto" experiments rely on trusting a person to make up his mind to press a button and then to rapidly choose to terminate his movement (Brass and Haggard 2007). A DP machine would finally allow one to perform true "veto" experiments. If it were possible not only to predict *when* a person is going to decide, but also *which specific option* he is going to take, one could ask him to change his mind and take the *opposite* option. It seems plausible that a person should be able to change his mind across a period as long as 7 s. However, there is a catch: How can one change one's mind if one doesn't even know what one has chosen in the first place? If it were one day realized, such a DP machine would be similarly useful in helping us realize the determination of our free decisions as an auto-cerebroscope (Feigl 1958) is in helping understand the relationship between our conscious thoughts and our brain activity.

Acknowledgments This work was funded by the Max Planck Society, the German Research Foundation and the Bernstein Computational Neuroscience Program of the German Federal Ministry of Education and Research. The author would like to thank Ida Momennejad for valuable comments on the manuscript.

References

Arieli A, Sterkin A, Grinvald A, Aertsen A (1996) Dynamics of ongoing activity: explanation of the large variability in evoked cortical responses. Science 273:1868–1871

Blankertz B, Dornhege G, Schäfer C, Krepki R, Kohlmorgen J, Müller KR, Kunzmann V, Losch F, Curio G (2003) Boosting bit rates and error detection for the classification of fast-paced motor commands based on single-trial EEG analysis. IEEE Trans Neural Syst Rehabil Eng 11:127–131

Brass M, Haggard P (2007) To do or not to do: the neural signature of self-control. J Neurosci 27:9141–9145

Brass M, Haggard P (2008) The what, when, whether model of intentional action. Neuroscientist 14:319–325

Breitmeyer BG (1985) Problems with the psychophysics of intention. Behav Brain Sci 8:539–540

Burgess PW, Quayle A, Frith CD (2001) Brain regions involved in prospective memory as determined by positron emission tomography. Neuropsychologia 39:545–555

Chiang T (2005) What's expected of us. Nature 436:150

Deiber MP, Passingham RE, Colebatch JG, Friston KJ, Nixon PD, Frackowiak RS (1991) Cortical areas and the selection of movement: a study with positron emission tomography. Exp Brain Res 84:393–402

Eccles JC (1982) The initiation of voluntary movements by the supplementary motor area. Arch Psychiatr Nervenkr 231:423–441

Eccles JC (1985) Mental summation: the timing of voluntary intentions by cortical activity. Behav Brain Sci 8:542–543

Feigl H (1958) The "mental" and the "physical". In: Feigl H, Scriven M, Maxwell G (eds) Minnesota studies in the philosophy of science: concepts, theories, and the mind-body problem, vol II. University of Minnesota Press, Minneapolis, MM
Fourneret P, Jeannerod M (1998) Limited conscious monitoring of motor performance in normal subjects. Neuropsychologia 36:1133–1140
Friston KJ, Holmes AP, Poline JB, Grasby PJ, Williams SC, Frackowiak RS, Turner R (1995) Analysis of fMRI time-series revisited. Neuroimage 2:45–53
Groll-Knapp E, Ganglberge JA, Haider M (1977) Voluntary movement-related slow potentials in cortex and thalamus in man. Progr Clin Neurophysiol 1:164–173
Haggard P, Eimer M (1999) On the relation between brain potentials and the awareness of voluntary movements. Exp Brain Res 126:128–133
Haynes JD, Rees G (2005) Predicting the orientation of invisible stimuli from activity in human primary visual cortex. Nat Neurosci 8:686–691
Haynes JD, Rees G (2006) Decoding mental states from brain activity in humans. Nat Rev Neurosci 7:523–34
Haynes JD, Sakai K, Rees G, Gilbert S, Frith C, Passingham RE (2007) Reading hidden intentions in the human brain. Curr Biol 17:323–328
Hume D (1777) An enquiry concerning human understanding. Reprinted as Harvard Classics, Collier & Son
Kamitani Y, Tong F (2005) Decoding the visual and subjective contents of the human brain. Nat Neurosci 8:679–685
Kornhuber HH, Deecke L (1965) Hirnpotentialänderungen bei Willkürbewegungen und passiben Bewegungen des Menschen: Bereitschaftspotential und reafferente Potentiale. Pflügers Arch Ges Phys 284:1–17
Kriegeskorte N, Goebel R, Bandettini P (2006) Information-based functional brain mapping. Proc Natl Acad Sci USA 103:3863–3868
Latto R (1985) Consciousness as an experimental variable: problems of definition, practice and interpretation. Behav Brain Sci 8:545–546
Libet B (1985) Unconscious cerebral initiative and the role of conscious will in voluntary action. Behav Brain Sci 8:529–566
Libet B, Gleason CA, Wright EW, Pearl DK (1983) Time of conscious intention to act in relation to onset of verebral activities (readiness-potential): the unconscious initiation of a freely voluntary act. Brain 106:623–642
Marks LE (1985) Toward a psychophysics of intention. Behav Brain Sci 8:547–548
Merikle PM, Cheesman J (1985) Conscious and unconscious processes: same or different? Behav Brain Sci 8:547–548
Moutoussis K, Zeki S (1997) Functional segregation and temporal hierarchy of the visual perceptive systems. Proc R Soc Lond B 264:1407–1414
Nickerson RS (2002) The production and perception of randomness. Psychol Rev 109:330–357
Ramnani N, Owen AM (2004) Anterior prefrontal cortex: insights into function from anatomy and neuroimaging. Nat Rev Neurosci 5:184–94
Ringo JL (1985) Timing volition: questions of what and when about W. Behav Brain Sci 8:550–551
Rollman GB (1985) Sensory events with variable central latencies provide inaccurate clocks. Behav Brain Sci 8:551–552
Soon CS, Brass M, Heinze HJ, Haynes JD (2008) Unconscious determinants of free decisions in the human brain. Nat Neurosci 11:543–545
Tanaka K (1997) Mechanisms of visual object recognition: monkey and human studies. Curr Opin Neurobiol 7:523–529
Trevena JA, Miller J (2002) Cortical movement preparation before and after a conscious decision to move. Conscious Cogn 1:162–190
Van de Grind W (2002) Physical, neural, and mental timing. Conscious Cogn 11:241–64
Wundt W (1904) Principles of physiological psychology. Vol. 2

Subliminal Motivation of the Human Brain

Mathias Pessiglione

Abstract Can our behavior be motivated by environmental signals that we are not aware of? In this chapter I cast light on this question with a series of experiments investigating whether the human brain can deal with the reward-predicting properties of visual stimuli that subjects cannot consciously perceive. The experimental paradigms designed for this purpose bring together procedures that have been used for decades in separate scientific fields: subliminal perception on one side and incentive motivation on the other. I first sketch a short history of methods and concepts used in these two fields, and then I present neuroimaging studies combining the two approaches to explore subliminal motivation in humans. Specifically, these studies demonstrate that the human brain is able to translate higher subliminal incentives into higher physical effort and to use subliminal cues that predict gambles outcomes to make profitable decisions. Subliminal motivation processes are underpinned by the ventral basal ganglia, under the influence of dopamine. I then suggest some clinical applications using subliminal paradigms to assess the integrity of basic motivation processes in patients while avoiding confounds related to distorted conscious judgments.

1 Introduction: A Brief History of Concepts

"Eat popcorn" is certainly one of the most famous subliminal messages. It was inserted between the images of a movie and presented so briefly that the audience could not see it. This happened in a cinema of the US West Coast in 1957. The owners of the cinema claimed that they had sold a lot of popcorn thanks to their subliminal manipulation. It appeared that they had invented a revolutionary technique: subliminal advertising. In fact, there had been no revolution; they just

---

M. Pessiglione (✉)
Institut du Cerveau et de la Moelle épinière, INSERM-CNRS, Université Pierre et Marie Curie, Paris, France
e-mail: mathias.pessiglione@gmail.com

S. Dehaene and Y. Christen (eds.), *Characterizing Consciousness: From Cognition to the Clinic?* Research and Perspectives in Neurosciences,
DOI 10.1007/978-3-642-18015-6_11,

invented their data, as they admitted a decade later. Nevertheless, a myth was born that has yielded a large-scale paranoia. Fertilized by Freudian theory, the idea was that one cannot fight against subliminal messages because they directly target the unconscious. This idea was so popular that commercial and electoral campaigns used subliminal messages to try to influence the choices of consumers or voters. No well-controlled study could prove such effects, however. Here, I revisit the phenomenon of subliminal motivation in a laboratory environment, merging techniques that have traditionally pertained to different fields of research: subliminal perception and incentive motivation.

2 Subliminal Perception

The empirical quest for a 'limen,' or threshold, below which perception is unconscious but still affects behavior is as old as experimental psychology. Pioneering studies showed that some capacity for discrimination can persist in the absence of consciousness. In those early days, conscious awareness was probed by introspection, with the subjects sometimes being the experimenters themselves (as in Pierce and Jastrow 1884). In one of these experiments, for instance, characters were presented at a distance, such that subjects reported seeing only a blurred spot. When forced to guess whether the characters were digits or letters, the subjects' performance was nonetheless better than chance (Sidis 1898). Several demonstrations of this kind, showing above-chance discrimination with no subjective awareness, were published during the first half of the twentieth century (see Adams 1957), until introspection was criticized as a valid method for assessing conscious awareness (Eriksen 1960). To properly demonstrate an absence of awareness, experimenters were instead required to show that an objective measure, such as discrimination performance, was at chance level. The ironic implication is that the same observation, above-chance discrimination, that was first interpreted as evidence for unconscious perception was now taken as an indicator of conscious awareness.

The dilemma is still unresolved: should we go for subjective or objective measures if we want to separate subconscious from conscious processes? Ideally we would prefer our demonstration not to depend on dubious reports about the subjective feeling of being conscious. As Eriksen (1960) pointed out, the same percept may be considered as conscious by some subjects but unconscious by other subjects, or by the same subjects under different instructions. In other words, introspective measures make consciousness dependent on a volatile confidence criterion. By contrast, an objective measure, such as the percentage of correct discriminations, seems much more reliable. The problem with stringent objective criteria is that they are likely to kill the very phenomenon, subconscious processing, that we intend to study. For example, with an objective criterion, one would dismiss such an interesting phenomenon as blindsight. This syndrome is commonly observed after damage to the primary visual cortex and is characterized by patients claiming they cannot see anything but showing impressive accuracy at discriminating between elementary visual patterns (Weiskrantz 1999). On one side, authors attached to objective criteria would conclude that blindsight patients have a conscious visual percept even if they fail to report it. On the other side, authors trusting subjective reports would draw the opposite conclusion: that discrimination performance has nothing to do with consciousness, since the two can be dissociated (Lau 2008). From a methodological viewpoint, subjective reports can be used to classify each stimulation as consciously perceived or not, and hence to monitor conscious awareness on a trial by trial basis (Baars 1988). On the contrary, discrimination performance is a statistical measure, which can only be assessed over a large number of trials, among which some occasional conscious perception may be missed. Indeed, demonstrating chance-level discrimination raises the formal issue of accepting a null hypothesis, which remains problematic because a negative test can mistake insufficient statistical power for an absence of effect (Cheesman and Merikle 1986).

In search for 'absolute subliminality,' some authors have formalized a dissociation procedure, which consists in showing that information about stimuli can be inaccessible for the conscious mind, as evidenced by a direct discrimination measure, but still available for some unconscious process indexed by an indirect measure (Reingold and Merikle 1988). This procedure has been largely implemented in masked priming studies, in which processing of a visible target is influenced by the prior presentation of a related invisible prime. For example, after being exposed to a masked word, say 'salt,' subjects were unable to state whether the word was present or not but they favored semantically related words, like 'pepper,' in a subsequent forced choice (Marcel 1983). These studies have been subjected to ferocious criticisms, however, such that subliminal semantic processing has been cyclically acclaimed and rejected, resulting at the end in substantial methodological improvement (see Kouider and Dehaene 2007). Among further methodological requirements, one is that an equal number of subliminal stimulations, in the same visual conditions, should be used for direct and indirect measures (Holender 1986). Another is that responses cannot be prepared in advance, to avoid direct motor specification, which might occur when using a small set of known stimuli (Abrams and Greenwald 2000; Damian 2001).

However, obtaining above-chance indirect measures simultaneously with chance-level direct measures does not necessarily equate with an effective subconscious processing in the absence of awareness. It would be the case if measures were both exclusive and exhaustive, meaning that the direct measure accounts for all and only conscious processing and the indirect measure all and only unconscious processing. These assumptions are hard to implement in practice, so it may be argued that the two measures are likely to represent a mix of conscious and unconscious processes. This is why some authors suggested that the only convincing evidence for subconscious processing is a qualitatively different behavioral effect compared to that obtained with conscious processing (Cheesman and Merikle 1986). For instance, the mere exposure effect involves subjects being flashed subliminal stimuli and then tested on both recognition and preference. Typically, in the recognition test subjects are unable to state which stimuli were previously presented, but in the preference judgment test they favor the previously presented stimuli (Kunst-Wilson and Zajonc 1980). Crucially, the effects are reversed when stimuli are explicitly and not subliminally presented, with above-chance recognition of old stimuli and preference for novel stimuli. A qualitative difference could in principle be found as well in neuroimaging data, if conscious and subconscious processing recruit different brain circuits. Unfortunately, most brain imaging studies have reported a quantitative difference, with the same regions being more activated when processing becomes conscious (see Marzi et al. 2004).

Nowadays, after a half century of intense controversy, subliminal perception can no longer be denied, whatever the criterion. But what representations exactly can be formed subconsciously is still a debated issue. An influential theory is that perception has two stages, with subconscious processing first for lower-level information and then conscious processing for higher-level representation (Baars 1988; Greenwald 1992; Dehaene and Naccache 2001). Although the frontier between lower and higher levels may still fluctuate, these authors agree that a large variety of short-lived information processing stages can remain subconscious, including those involved in semantic representations. Some other stages may, however, need to be conscious, notably the strategic processing that would develop under volitional control and apply to longer time scales. Curiously, there was little investigation on whether and how the brain can process the reward-predicting properties of subliminal stimuli before our own publications (Pessiglione et al. 2007, 2008). One earlier attempt to motivate people with subliminal message was the famous "eat popcorn" flashed on a movie screen on the US west coast in 1957. The authors first claimed that the subliminal manipulation had boosted their sales but later admitted this was a fake (see Pratkanis 1992). Despite the fact that no well-controlled study could prove any effect on consumer behavior, the idea that you cannot defend yourself against subliminal advertising, because it directly targets your unconscious, became highly popular. Our aim was to revisit the concept of subliminal motivation by applying a visual masking technique to paradigms developed for conditioning studies in animals. The general plan was to degrade the visibility of reward-predicting stimuli and look for dissociations between direct indicators, both objective and subjective, of conscious awareness, and for indirect measures of incentive motivation.

3 Incentive Motivation

Motivation is not an easy concept to investigate empirically, because it belongs to common language and accepts vague and various meanings. Several operational definitions of motivation were proposed in the course of the last century (see Berridge 2004). During the behaviorist era, motivation was first seen as a useless concept because stimulus-response relationships were believed to exhaustively describe the observations. There is indeed a threat of circularity with motivational explanation: when an animal executes an action, it adds nothing to say that the animal had a motivation to execute the action. Later on, motivation was employed as a convenient intervening variable between experimental manipulation and behavioral observation (see Miller 1971). The issue was that various manipulations can produce the same spectrum of observations; for instance, water deprivation and heat dehydration may result in more water drunk or more work for a sip. Instead of making all possible stimulus-response links, which would rapidly lead to a combinatorial explosion, it is parsimonious to link the above-cited manipulations to thirst and then link thirst to the behavioral observations. At this period (until 1960s), motivation was equated with homeostatic drive, assuming that behavior resulted from the need to keep some internal variables, such as glucose plasma level, at given set points. Drives like thirst or hunger would then develop from an error signal and trigger compensatory behavior to minimize the error. Thus, motivational explanations were formulated in terms of drive reduction, which may have reasonable face validity for food - or water-directed behavior but much less for sex or dominance. Research on the biological correlates of motivation has, nonetheless, long been focused on identifying set points and error signals that would trigger drive-reduction behavior.

Several difficulties led most authors to abandon the drive-reduction account of motivation. It was first argued that a transiently stable balance between opposing forces can give the illusion of homeostatic regulation. For instance, body weight may have no set point at all, but only fluctuating levels resulting from a balance between opponent systems. According to (Bolles 1980), the fact that obesity rates have recently risen does not reflect a change in internal set points but in external availability and palability of food, which would favor eating behavior. Thus, as other authors would put it, hedonic aspects of rewards may not be negligible in explaining behavior (Pfaffmann 1960; Young 1966; Cabanac 1992). Further evidence was drawn from a patient whose esophagus was permanently damaged and who was fed through gastric fistula. Surprisingly, the patient insisted on chewing food at meals before placing it in his stomach (Wolf and Wolff 1943). This situation was reproduced in animals to show that satisfying appetite is not merely a matter of drive reduction (Miller and Kessen 1952). Another line of evidence came from self-stimulation studies in animals with implanted electrodes. If the drive-reduction theory were true, stimulation of a brain site that triggers eating behavior should be aversive, meaning that rats would want to stop it. The prediction turned out to be false: when given the opportunity, the rats pressed levers to prolong the stimulation, just as they pressed levers to obtain food, such that eating electrodes were also reward electrodes (Valenstein 1976).

The concept of incentive motivation was built on earlier reflections about the necessary features of a truly motivated behavior. It was first noted that all motivated behavior could be divided into appetitive and consummatory phases (Sherrington 1906). The appetitive phase involves flexible approach behavior directed toward a goal, whereas the consummatory phase consists in stereotyped behavioral patterns elicited when the goal is within reach, such as chewing food, aggressive biting or sexual copulation. Teitelbaum (1966) pointed to the flexibility of the appetitive phase as an essential feature of motivated behavior, including the possibility of learning a new operant response or inferring a new strategy to obtain the goal. Epstein (1982) added two criteria, goal expectation and affective reaction, that are supposed to distinguish between truly motivated behavior and that of machine learning algorithms. The most basic reinforcement learning algorithm would be something like "increase response frequency when you get a reward." Such an algorithm cannot account for some behavioral markers of motivation such as autoshaping or incentive contrast (Crespi 1942; Williams and Williams 1969; Jenkins and Moore 1973). These effects are obtained when the reward used to reinforce an operant response is suddenly increased: certain behavioral measures that are irrelevant to obtaining the reward, such as the speed of approach response or the rate of anticipatory licking, will change on the very next trial before any reinforcement can take place. These effects are usually explained by the fact that animals are expecting a reward and not just activating context-action associations. Also, by contrast with learning machines, the affective manifestations (including somatic, autonomic or hormonal markers) of the appetitive phase indicate that behavior is really motivated towards hedonic goals. Furthermore, Balleine and Dickinson (1998) emphasized that not only the expected reward but also the contingency linking the behavior to the reward has to be represented during the appetitive phase This can be tested in extinction paradigms following devaluation of the reward, by satiation for instance. If the behavior is reduced, then it can be qualified as goal directed, meaning dependent on updated reward expectation. Otherwise, it would be called a habit, which is mere activation of context-action links without any goal representation.

Thus, contrary to the drive-reduction theory, where everything starts with homeostasis, incentive motivation theory makes things start with the hedonic experience of rewards. In the latter view, what motivates people is learned expectation of hedonic rewards (Bolles 1972). This definition fits well with common intuition: when we question the motivation, or the motive, for an action, we basically wonder what expected reward the action is aimed at. More specifically, it has been assumed that the value of hedonic reward experience is transferred through associative learning to contingent cues, which thereby acquire motivational properties (Bindra 1974). Of course, the proponents of incentive motivation theory would not claim that physiological states, such as hunger and satiation, play no role in motivating behavior. They propose instead that physiological states could magnify or attenuate the incentive/hedonic value of reward predictive cues (Toates 1986). According to Berridge and Robinson (1998), the predictive cues become both liked and wanted, forming a gradient along which animals may move, from the most distal to the most proximal, up to reaching the reward. In their terminology, 'liking' is an affective state reflecting the hedonic properties of the cue, whereas 'wanting,' also called 'incentive salience,' reflects the motivational value of the cue, which may attract attention and behavior. In normal situations, 'liking' and 'wanting' usually go together, but they can be dissociated following certain brain modifications; for instance, drug addicts may still want to take substances that they do not particularly like anymore.

Associative learning is thus central to incentive motivation theory, which can be seen as a cognitive interpretation of the animal conditioning literature that had been constituted during the behaviorist era, when cognitive representations were banished. In classical or Pavlovian conditioning, an arbitrary cue (say a bell ring) is repeatedly followed by a rewarding outcome (say a sausage), until it shows the capacity to elicit a vegetative manifestation (say salivation) that is normally observed in response to the reward (Pavlov 1927). In operant or instrumental conditioning, access to reward (say a food pellet) is dependent on a behavioral response (say lever press), the frequency of which increases across trials according to the law of effect (Thorndike 1911). Thus, Pavlovian learning involves building cue-outcome associations and instrumental learning involves cue-response-outcome associations. Both types of learning depend on two main factors: contiguity (short delays between reward and cue or response) and contingency (more reward following specific cue or response). It is generally accepted that, in both cases, the learning rate is commensurate with a reward prediction error, which is the difference between actual and predicted reward (Rescorla and Wagner 1972). Basically, this rule says that surprise may serve as a teaching signal: the more surprised you are the more you learn about your environment. Note that the surprise has a sign: a positive prediction error (more reward than expected) will reinforce the associations between cues, responses and outcomes, whereas negative ones (less reward than expected) will weaken these associations. The crucial role of prediction errors is well illustrated in blocking paradigms, where a reward that is already fully predicted by a first cue fails to be associated with other cues (Kamin 1969). The Rescorla–Wagner learning rule has been developed and implemented in various models, providing a computational account of the incentive salience concept, which would reflect reward prediction for a given cue (Schultz et al. 1997; McClure et al. 2003). If we imagine a sequential chain of cues leading to reward, the incentive salience gradient would be learned from a teaching signal, equivalent to the difference of reward prediction between two successive cues, or two time points (the so-called 'temporal difference error').

Beyond the acquisition of incentive salience, we were interested in its effects on behavior, that is, in the motivational processes occurring during the appetitive phase of goal-directed action. Outside the learning context, motivational effects can be divided into two types: energizing and directing behavior. Energizing behavior can be assessed as the amount of effort, for instance the number of lever presses, exerted by an animal to obtain a given reward (Walton et al. 2006). Directing behavior can be assessed as the alternative favored by the animal in choice situations, for instance, in discounting paradigms, where a small reward is compared to a larger reward associated with higher effort, longer delay or lower probability (Cardinal 2006). Working with humans, we chose to use monetary rewards, which offer several advantages: they are easy to quantify, they are not prone to satiation, and they provide a natural counterpart, losing money as opposed to winning money, to contrast punishment avoidance with reward seeking. The paradigms detailed below were originally designed for neuroimaging studies, in which we analyzed the BOLD signal to identify key brain areas and the skin conductance response as an index of affective reactions (Pessiglione et al. 2007, 2008). Here, we focus on behavioral measures and develop some clinical applications to assess patients in whom motivation is either deficient (as in cases of apathy) or hard to control (as in cases of impulsivity).

4 Experimental Studies

4.1 Energizing Behavior: Can We Subliminally Drive People to Try Harder?

A first series of experiments was designed to examine whether monetary incentives can subconsciously influence motor activation. We manipulated both the visibility and the amount of money presented in the different trials: subjects were flashed various coin images between two visual masks, with various durations. In a visual discrimination task, subjects were asked to report which incentive had been flashed and whether they saw it or had just been guessing. We thus observed two direct indicators of stimulus awareness: percentage of correct responses (objective measure) and percentage of seen responses (subjective measure). In the incentive force task, subjects were asked to squeeze a handgrip and explained that they would be allowed to keep a fraction of the monetary incentive proportional to the force produced. Handgrip force was thus taken as an indirect measure reflecting incentive motivation. Subjects were given online visual feedback on the force exerted, visualized as a cursor moving up and down, and a cumulative total of the money won at the end of every trial. We used two criteria to select sessions where subjects were not consciously aware of the monetary incentive: objective measure at chance level and subjective measure at zero. In both cases, subjects still exerted more force for higher incentives (Pessiglione et al. 2007). This effect of incentive motivation was specifically underpinned by bilateral activation of a basal forebrain region known to be involved in reward processing (Robbins and Everitt 1996; Pecina et al. 2006). The same region was activated regardless of stimulus visibility, such that conscious awareness made no qualitative difference (no separate circuits for conscious and subconscious motivation) but made a quantitative difference (more activation when motivation becomes conscious). The energizing effect of monetary incentives was accompanied by an increase in skin conductance response, suggesting an emotional participation in addition to physical effort. The two measures, skin conductance and grip force, might reflect, respectively, the liking and wanting components of incentive motivation defined by Berridge and Robinson (2003).

Using a similar task but with supraliminal stimuli only, we showed that the two components can be dissociated following bilateral lesions of the basal ganglia (Schmidt et al. 2008): in these patients, the skin conductance is still responsive to the amount of money at stake (liking is intact), but the grip force is no longer adapted (wanting is impaired). The syndrome exhibited by these patients is a severe apathy called 'auto-activation deficit' (Laplane and Dubois 2001). These patients would spontaneously do nothing even if they can perform even complex tasks under external solicitation, indicating that their motor and cognitive abilities are intact. Indeed, in our experiment we showed that auto-activation deficit (AAD) patients could produce different levels of force on demand, but that they produced the same force whatever the monetary incentive was. Thus, their motor ability to modulate their effort is preserved but remains insensitive to potential rewards. Because skin conductance responses were graduated in proportion to the monetary incentives, we conclude that basal ganglia damage induced a disconnection between the affective and the motor responses to expected rewards. In other words, the brain mechanism that translates higher incentives into higher effort was impaired in patients with AAD.

This mechanism appeared to also be impaired in patients with major depression, who produced no more effort for higher incentives (Cléry-Melin et al., submitted for publication). The difference between AAD and depressed patients was found in the skin conductance response: it was sensitive to monetary incentives in AAD but not in depressed patients. Hence, major depression alters both the affective and motor responses to potential rewards. However, subjective ratings of perceived efforts (how hard they tried) were enhanced in depressed patients whereas they were decreased in healthy controls. This result indicates that incentive motivation made the same task feel easier in healthy participants but harder in depressed patients. Thus, contrary to the common impression that depressed patients make no effort to improve their lot, they might be at the maximum they can withstand even for seemingly easy work. The result also indicates that patients took into account the money at stake in their judgment about their own performance. Instead of a primary failure to energize behavior in response to potential rewards, a more sophisticated interpretation may be advanced. Patients would guess that they were expected to perform better but felt unable to fulfill this expectation, which would both increase fatigue sensation and prevent them from engaging more effort. In this case, the incentive motivation deficit would be secondary to a distortion of performance judgments. The subliminal motivation paradigm may help disentangle these alternative hypotheses: if patients do produce more force following subliminal incentives, then the deficit observed with supraliminal incentives was due to a distortion of performance judgments.

Taken together, our results provide strong evidence for the possibility of subliminal motivation: invisible stimuli predicting more reward have the capacity to energize behavior. This motivational interpretation was also supported by the other measures recorded during the original neuroimaging experiment, notably the specific activation of limbic basal ganglia circuits (and not motor or associative circuits) in the contrast between incentives, and the higher skin conductance response for higher incentives, revealing involvement of affective processes in addition to putative cold decision-making and motor execution. The concept of subliminal motivation received further empirical support in a publication reporting that subliminal words expressing effort-related concepts can increase force production (Aarts et al. 2008). However, subliminal motivation effects were observed here with coin images, when they had their habitual meaning, the reward properties of which have been well established. In the next section I address the issue of whether neutral subliminal stimuli can acquire reward properties so as to influence choices.

4.2 Directing Behavior: Can We Subliminally Drive People to Gamble?

A second series of experiments was designed to assess whether instrumental learning can occur subconsciously. Initial steps had already been taken towards this demonstration, notably with the so-called implicit learning concept, meaning that behavioral responses can be adapted to the statistical structure of stimuli that fails to be reported explicitly (Knowlton et al. 1996; Destrebecqz and Cleeremans 2001; Bayley et al. 2005). In implicit learning tasks using artificial grammar, serial reaction time or probabilistic classification, authors have claimed that subjects can achieve good acquisition without explicit knowledge of the task structure. However, methods for assessing awareness of statistical contingencies have been criticized, principally on the issue that failure to answer retrospective questions about task structure does not prove any absence of awareness at the time of performing the task. In other words, questions were very demanding in terms of memory and subjects may have simply forgotten what they had been aware of (Wilkinson and Shanks 2004; Lagnado et al. 2006). Thus, to formally test whether instrumental conditioning can occur without awareness, we took a more stringent approach: masking the cues, so that they remained unperceived. Subliminal conditioning studies in humans had so far been restricted to Pavlovian paradigms such as fear conditioning (Clark and Squire 1998; Knight et al. 2003; Olsson and Phelps 2004), where subliminal stimuli (like unseen faces) are paired with unpleasant events (like white noise) to increase autonomic responses (like skin conductance). To our knowledge, subliminal instrumental conditioning, where decision making would be biased by unperceived cues predicting rewards or punishments, had never been demonstrated.

We used the same abstract cues, which were letters taken from a medieval font, and the same masking procedure to set up our visual discrimination and instrumental conditioning tasks (Fig. 1a). In the visual discrimination task, two masked cues were successively displayed on the screen, and subjects had to report whether or not they perceived a difference between the two stimulations. We reasoned that if subjects are unable to correctly discriminate between the masked cues, then they are also unable to build conscious representations of cue-outcome associations. We chose not to use a recognition test because identifying the cues would not be necessary to build conscious associations with the outcomes; discriminating them would indeed be sufficient. The procedure also has the advantage of not showing

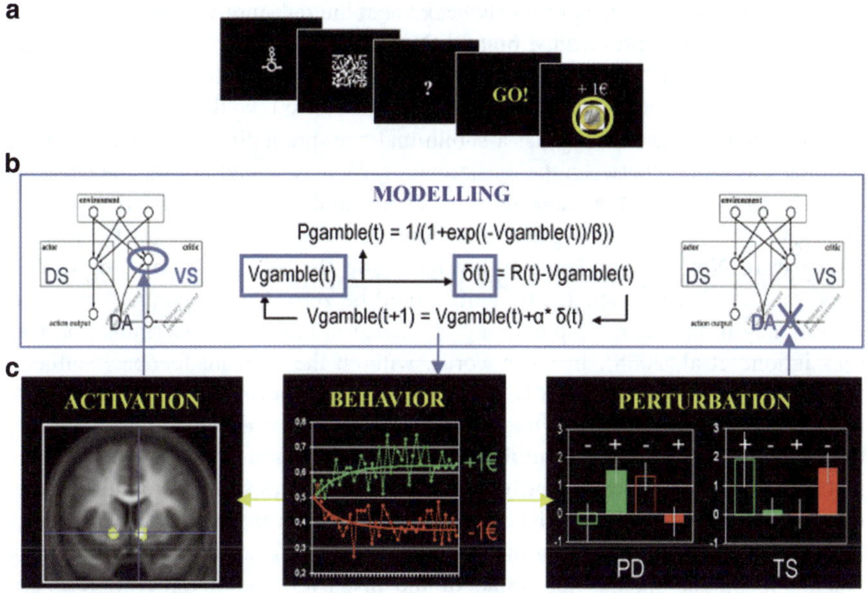

Fig. 1 Subliminal learning: overview of methods and results. (**a**) The subliminal instrumental conditioning task. Successive screenshots displayed during a given trial are shown from *left* to *right*, with durations in milliseconds. After seeing a masked contextual cue flashed on a computer screen, subjects choose to gamble or not, and subsequently observe the outcome. In this example, 'Go' appears on the screen because the subject has opted for gambling, following the cue associated with reward (winning 1€). (**b**) Computational model. The model is sketched as an actor-critic architecture (*left* and *right*) and mathematically described as a Q-learning algorithm (*middle*). From the same contextual cues, the critic learns to predict rewards and the actor learns to select actions. For both the actor and the critic, the teaching signal is the reward prediction error (delta), that is the difference between the predicted reward (Vgamble) and the actual reward (R). To compute the probability of gambling choice (Pgamble), the model uses a softmax equation, which is a sigmoid function of the learned value of gambling (Vgamble). To update this value, the model uses the Rescorla and Wagner equation, which specifies that changes between two successive trials (t+1 versus t) are proportional to reward prediction errors. (**c**) Main results. Middle panel: learning curves (data points) show that healthy participants learned to gamble following reward cues and to avoid gambling following punishment cues. *Continuous lines* show the gambling probabilities generated by the model after optimization of the free parameters (alpha in the learning rule and beta in the choice rule). *Left panel*: neuroimaging data show that ventral striatum (VS) activity encoded the theoretical values of gambling generated by the model. The VS might therefore play the role of the critic, learning the values attached to the different cues. We also suggest that the dorsal striatum (DS) might play the role of the actor, selecting rewarded actions. *Right panel*: dopamine-related medications (+ means 'on state') have opposite effects on learning performance in patients with Parkinson's disease (PD) and Tourette's syndrome (TS). Dopamine (DA) release might therefore encode reward prediction errors, boosting approach learning following positive outcomes but impairing avoidance learning following negative outcomes

the cues unmasked, so that, by the end of the experiment, subjects had no idea what the cues looked like. In the instrumental conditioning task, subjects had to decide whether or not to make a risky response (that is, to gamble), which could be pressing or not pressing a button. Feedback about the response was displayed on the screen: 'Go!' if they pressed the button, 'No!' if they did not. Then they observed the outcome, which was neutral following a safe response and either rewarding (monetary gain), neutral or punishing (monetary loss) in the event of a risky response. Subjects were told that a subliminal cue, predicting the outcome of the gamble, was hidden between the mask images. As they would not see the cues, we encouraged them to follow their intuition, choosing the risky response if they had a feeling they were in a winning trial and choosing a safe response if they felt it was a losing trial. Note that if subjects always made the same response, or if they performed at chance, their final payoff would be zero. On the contrary, subjects won money in this task, despite being at chance in the visual discrimination task (Pessiglione et al. 2008). In other words, without the relevant feedback subjects stayed at chance, but with the help of rewards and punishments they learned to differentiate their responses to the various masked cues (Fig. 1c, middle).

To model instrumental conditioning, we implemented a standard Q-learning algorithm, with inputs from individual histories of cues, choices and outcomes (Fig. 1b, middle). On every trial, the model estimates the likelihood of the risky response from the value of the displayed cue. If the risky response was actually taken, the model updates the value of the displayed cue in proportion to the prediction error, according to the Rescorla and Wagner rule. The parameters of the model were optimized such that likelihoods of risky responses provided the best fit of subjects' actual responses across conditioning sessions. The theoretical values of subliminal cues generated by this optimized algorithm were then used as regressors for analysis of brain imaging data. Cue value correlated with activity in the ventral striatum, a component of limbic basal ganglia circuits (Fig. 1c, left). The same region was activated in a previous study using supraliminal cues such that, again, conscious awareness made no qualitative difference in the brain circuits recruited to perform the task. Interestingly, the striatum has been identified as a major player in implicit/procedural (but not explicit/declarative) learning (Hikosaka et al. 1999; Packard and Knowlton 2002; Graybiel 2005). In our case, there was no correlation between ventral striatum activity and behavioral performance, indicating that this brain region equally learned the cue-outcome associations in all subjects, although not all subjects managed to bias their choices appropriately. The implication is that some learning occurred at the neuronal level even in subjects who were playing at chance in the instrumental conditioning task.

We then turned to patients to investigate the role of dopamine signalling in subliminal instrumental learning. Single-unit recording studies in monkeys have shown that reward prediction errors are encoded by dopamine neurons (Schultz et al. 1997). Dopamine transmission is the target of many medications used in neurological or psychiatric conditions. For instance, dopamine enhancers (dopamine precursors or receptor agonists) are used to alleviate hypokinetic symptoms (poor movements) of Parkinson's disease (PD), whereas dopamine blockers (antagonists of dopamine receptors) are used to reduce hyperkinetic symptoms (motor and vocal tics) in Tourette's syndrome (TS). We administered our subliminal conditioning paradigm to PD and TS patients while they were taking ('on state') or not taking ('off state') their medication (Fig. 1c, right). We found that dopamine enhancers boosted reward learning but impaired punishment learning in PD patients (Palminteri et al. 2009). A mirror double dissociation was found in TS patients, with dopamine blockers favoring punishment learning but impairing reward learning. These results show that dopamine-related drugs can provide a bias on subconscious learning and hence on gambling decisions. A similar double dissociation had been found in PD patients using supraliminal cues (Frank et al. 2004). To account for this result, Michael Frank suggested a model of basal ganglia circuits in which positive prediction errors (unexpected rewards) result in dopamine releases that reinforce approach pathways, whereas negative prediction errors (unexpected punishments) result in dopamine dips that reinforce avoidance pathways. This model would also explain our data: enhancing dopamine release would favor reward learning, whereas blocking dopamine release would favor punishment learning. The same mechanisms could operate whether subjects are aware or not of the cues that predict gambling outcomes.

5 Conclusion

Applying the standard masking procedures used in subliminal perception studies, we rendered stimuli inaccessible to the conscious mind. Following the definition of Dehaene et al. (2006), these stimuli can be said to be subliminal because they were attended but not reportable. When asked a direct question about the stimuli, subjects reported not seeing anything (subjective criterion) and their discrimination performance was at chance level (objective criterion). Adapting to humans some paradigms borrowed from animal conditioning literature, we contrasted these direct measures with indirect measures indexing motivational effects: first the amount of effort expended to win a subliminal incentive, and second the propensity to take a risky response after a subliminal abstract cue. Although subjects were unable to discriminate the relevant dimensions of subliminal stimuli, they nonetheless exhibited motivational effects. Namely, subliminal stimuli had the capacity to energize behavior, to acquire rewarding values through associations with outcomes, and to bias choices towards advantageous gambling. This is not saying that conscious awareness plays no role in motivation processes, as consciously visible stimuli always yielded quantitatively greater effects compared with subliminal ones.

Although it has not been demonstrated empirically before, that motivation can be subconscious may not be surprising for the general public. In particular, Freudian theory and subliminal advertising have both popularized the view that we can be motivated against our conscious will. Some differences from the concept of subliminal incentive motivation that we have evidenced here are worth pointing out. A notable difference is that motivational impulses in Freudian theory come from the inside, not from the environment. Also, subliminal stimuli never influenced behavior in a way that would not be desirable for the subjects. In other words, had they been aware of the stimuli, subjects would have behaved in a similar and even amplified way. Thus, there seems to be no value for advertisers in masking their messages; they would just reduce their impact on consumer behavior. How long the effects of subliminal advertising would last is not known. In our original study (Pessiglione et al. 2008), we showed that subliminal instrumental conditioning could bias the preferences even after unmasking the cues. Subjects preferred the cues that had been paired with rewards compared to those associated with punishments. However, this preference judgment test was done about 10 minutes after the conditioning sessions, and nobody knows whether or not the effect would survive an hour.

In conclusion, we have gathered a set of empirical evidence suggesting that incentive motivation can be triggered by stimuli that cannot be reported. We believe that the paradigms developed here may serve to assess patients in whom motivation is deficient (apathy, depression) or hard to control (impulsivity, compulsivity, addiction). Using subliminal presentations would avoid confounds relative to metacognitive processes, which might be dysfunctional in patients. In major depression, for instance, it might be that, when patients remain unaware of the rewards at stake, they avoid maladaptive conscious expectations about their own performance and hence engage in more constructive behavior.

References

Aarts H, Custers R, Marien H (2008) Preparing and motivating behavior outside of awareness. Science 319:1639

Abrams RL, Greenwald AG (2000) Parts outweigh the whole (word) in unconscious analysis of meaning. Psychol Sci 11:118–124

Adams JK (1957) Laboratory studies of behavior without awareness. Psychol Bull 54:383–405

Baars BJ (1988) A cognitive theory of consciousness. Cambridge University Press, New York

Balleine BW, Dickinson A (1998) Goal-directed instrumental action: contingency and incentive learning and their cortical substrates. Neuropharmacology 37:407–419

Bayley PJ, Frascino JC, Squire LR (2005) Robust habit learning in the absence of awareness and independent of the medial temporal lobe. Nature 436:550–553

Berridge KC (2004) Motivation concepts in behavioral neuroscience. Physiol Behav 81:179–209

Berridge KC, Robinson TE (1998) What is the role of dopamine in reward: hedonic impact, reward learning, or incentive salience? Brain Res Brain Res Rev 28:309–369

Berridge KC, Robinson TE (2003) Parsing reward. Trends Neurosci 26:507–513

Bindra D (1974) A motivational view of learning, performance, and behavior modification. Psychol Rev 81:199–213

Bolles RC (1972) Reinforcement, expectancy, and learning. Psychol Rev 79:394–409

Bolles RW (1980) Some functionalistic thoughts about regulation. In: Toates TW, Halliday TW (eds) Analysis of motivational processes. Academic, New York, pp 63–75

Cabanac M (1992) Pleasure: the common currency. J Theor Biol 155:173–200

Cardinal RN (2006) Neural systems implicated in delayed and probabilistic reinforcement. Neural Netw 19:1277–1301

Cheesman J, Merikle PM (1986) Distinguishing conscious from unconscious perceptual processes. Can J Psychol 40:343–367

Clark RE, Squire LR (1998) Classical conditioning and brain systems: the role of awareness. Science 280:77–81
Crespi LP (1942) Quantitative variation of incentive and performance in the white rat. Am J Psychol 55:467–517
Damian MF (2001) Congruity effects evoked by subliminally presented primes: automaticity rather than semantic processing. J Exp Psychol Hum Percept Perform 27:154–165
Dehaene S, Naccache L (2001) Towards a cognitive neuroscience of consciousness: basic evidence and a workspace framework. Cognition 79:1–37
Dehaene S, Changeux JP, Naccache L, Sackur J, Sergent C (2006) Conscious, preconscious, and subliminal processing: a testable taxonomy. Trends Cogn Sci 10:204–211
Destrebecqz A, Cleeremans A (2001) Can sequence learning be implicit? New evidence with the process dissociation procedure. Psychon Bull Rev 8:343–350
Epstein AN (1982) The physiology of thirst. In: Pfaff DW (ed) Physiological mechanisms of motivation. Springer, New York, pp 25–55
Eriksen CW (1960) Discrimination and learning without awareness: a methodological survey and evaluation. Psychol Rev 67:279–300
Frank MJ, Seeberger LC, O'Reilly RC (2004) By carrot or by stick: cognitive reinforcement learning in parkinsonism. Science 306:1940–1943
Graybiel AM (2005) The basal ganglia: learning new tricks and loving it. Curr Opin Neurobiol 15:638–644
Greenwald AG (1992) New look 3 Unconscious cognition reclaimed. Am Psychol 47:766–779
Hikosaka O, Nakahara H, Rand MK, Sakai K, Lu X, Nakamura K, Miyachi S, Doya K (1999) Parallel neural networks for learning sequential procedures. Trends Neurosci 22:464–471
Holender D (1986) Semantic activation without conscious identification in dichotic listening, parafoveal vision, and visual masking: a survey and appraisal. Behav Brain Sci 9:1–23
Jenkins HM, Moore BR (1973) The form of the auto-shaped response with food or water reinforcers. J Exp Anal Behav 20:163–181
Kamin LJ (1969) Selective association and conditioning. In: Mackintosh NJ, Honig WK (eds) Fundamental issues in instrumental learning. Dalhousie University Press, Halifax, NS, pp 42–64
Knight DC, Nguyen HT, Bandettini PA (2003) Expression of conditional fear with and without awareness. Proc Natl Acad Sci USA 100:15280–15283
Knowlton BJ, Mangels JA, Squire LR (1996) A neostriatal habit learning system in humans. Science 273:1399–1402
Kouider S, Dehaene S (2007) Levels of processing during non-conscious perception: a critical review of visual masking. Philos Trans R Soc Lond B Biol Sci 362:857–875
Kunst-Wilson WR, Zajonc RB (1980) Affective discrimination of stimuli that cannot be recognized. Science 207:557–558
Lagnado DA, Newell BR, Kahan S, Shanks DR (2006) Insight and strategy in multiple-cue learning. J Exp Psychol Gen 135:162–183
Laplane D, Dubois B (2001) Auto-activation deficit: a basal ganglia related syndrome. Mov Disord 16:810–814
Lau HC (2008) A higher order Bayesian decision theory of consciousness. Prog Brain Res 168:35–48
Marcel AJ (1983) Conscious and unconscious perception: experiments on visual masking and word recognition. Cogn Psychol 15:197–237
Marzi CA, Minelli A, Savazzi S (2004) Is blindsight in normals akin to blindsight following brain damage? Prog Brain Res 144:295–303
McClure SM, Daw ND, Montague PR (2003) A computational substrate for incentive salience. Trends Neurosci 26:423–428
Miller NE (1971) Neal E. Miller: selected papers. Aldine Atherton, Chicago, IL
Miller NE, Kessen ML (1952) Reward effects of food via stomach fistula compared with those of food via mouth. J Comp Physiol Psychol 45:555–564
Olsson A, Phelps EA (2004) Learned fear of "unseen" faces after Pavlovian, observational, and instructed fear. Psychol Sci 15:822–828

Packard MG, Knowlton BJ (2002) Learning and memory functions of the basal ganglia. Annu Rev Neurosci 25:563–593
Palminteri S, Lebreton M, Worbe Y, Grabli D, Hartmann A, Pessiglione M (2009) Pharmacological modulation of subliminal learning in Parkinson's and Tourette's syndromes. Proc Natl Acad Sci USA 106:19179–19184
Pavlov PI (1927) Conditioned reflexes. Oxford University Press, London
Pecina S, Smith KS, Berridge KC (2006) Hedonic hot spots in the brain. Neuroscientist 12:500–511
Pessiglione M, Schmidt L, Draganski B, Kalisch R, Lau H, Dolan RJ, Frith CD (2007) How the brain translates money into force: a neuroimaging study of subliminal motivation. Science 316:904–906
Pessiglione M, Petrovic P, Daunizeau J, Palminteri S, Dolan RJ, Frith CD (2008) Subliminal instrumental conditioning demonstrated in the human brain. Neuron 59:561–567
Pfaffmann C (1960) The pleasures of sensation. Psychol Rev 67:253–268
Pierce CS, Jastrow J (1884) On small differences in sensation. Mem Natl Acad Sci 3:75–83
Pratkanis AR (1992) The cargo-cult science of subliminal persuasion. Skept Inq 16:260–272
Reingold EM, Merikle PM (1988) Using direct and indirect measures to study perception without awareness. Percept Psychophys 44:563–575
Rescorla RA, Wagner AR (1972) A theory of Pavlovian conditioning: variations in the effectiveness of reinforcement and nonreinforcement. In: Black AH, Prokasy WF (eds) Classical conditioning II: current research and theory. Appleton-Century-Crofts, New York, pp 64–99
Robbins TW, Everitt BJ (1996) Neurobehavioral mechanisms of reward and motivation. Curr Opin Neurobiol 6:228–236
Schmidt L, d'arc BF, Lafargue G, Galanaud D, Czernecki V, Grabli D, Schupbach M, Hartmann A, Levy R, Dubois B, Pessiglione M (2008) Disconnecting force from money: effects of basal ganglia damage on incentive motivation. Brain 131:1303–1310
Schultz W, Dayan P, Montague PR (1997) A neural substrate of prediction and reward. Science 275:1593–1599
Sherrington CS (1906) The integrative action of the nervous system. C. Scribner's Sons, New York
Sidis B (1898) The psychology of suggestion. Appleton, New York
Teitelbaum P (1966) The use of operant methods in the assessment and control of motivational states. In: Honig WK (ed) Operant behavior: areas of research and application. Appleton-Century-Crofts, New York, pp 565–608
Thorndike EL (1911) Animal intelligence: experimental studies. Macmillan, New York
Toates F (1986) Motivational systems. Cambridge University Press, Cambridge
Valenstein ES (1976) The interpretation of behavior evoked by brain stimulation. In: Wauquier A, Rolls ET (eds) Brain-stimulation reward. Elsevier, New York, pp 557–575
Walton ME, Kennerley SW, Bannerman DM, Phillips PE, Rushworth MF (2006) Weighing up the benefits of work: behavioral and neural analyses of effort-related decision making. Neural Netw 19:1302–1314
Weiskrantz L (1999) Consciousness lost and found. Oxford University Press, Oxford
Wilkinson L, Shanks DR (2004) Intentional control and implicit sequence learning. J Exp Psychol Learn Mem Cogn 30:354–369
Williams DR, Williams H (1969) Auto-maintenance in the pigeon: sustained pecking despite contingent non-reinforcement. J Exp Anal Behav 12:511–520
Wolf S, Wolff HG (1943) Human gastric function, an experimental study of a man and his stomach. Oxford University Press, London
Young PT (1966) Hedonic organization and regulation of behavior. Psychol Rev 73:59–86

From Conscious Motor Intention to Movement Awareness

E.A. Fridman, M. Desmurget, and A. Sirigu

Abstract Willed actions are characterized by the dual subjective experiences of intention and awareness. Recent evidence suggests that these two experiences are not independent but interrelated at both the phenomenological and neurophysiological levels. Phenomenological evidence shows that the motor system is mainly aware of its intention, i.e., of what it wants to do. Neurophysiological observations indicate that movement intention and motor awareness share a common neural substrate in the posterior parietal cortex. These commonalities can be understood under the assumption that awareness relies on predictive (forward) mechanisms making use of our initial intentions.

1 Introduction

How do conscious intentions emerge and how do we become aware of what we are doing? For several centuries, these simple questions have been a major subject of investigation for theologians and philosophers like Saint Augustine, Descartes, Spinoza and Bergson. However, at the beginning of the twentieth century, Watson suggested that intention and awareness were not possible objects for scientific investigation. This view held for a long time and only eroded recently in the face of neuropsychological evidence suggesting that intentionality and consciousness were not spiritual traits but emerging properties of neural activities. When the brain is damaged, our capacity to generate conscious intentions to act can be severely impaired (Haggard 2008; Desmurget and Sirigu 2009) and our ability to be aware of our motor responses can be dramatically altered (Frith et al. 2000). Regarding intentionality, for instance, some patients can become spectators to alien movements

A. Sirigu (✉)
Centre de Neuroscience Cognitive, UMR 5229, CNRS, Bron, France
and
Université Claude Bernard, Lyon 1, Lyon, France
e-mail: sirigu@isc.cnrs.fr

S. Dehaene and Y. Christen (eds.), *Characterizing Consciousness: From Cognition to the Clinic?* Research and Perspectives in Neurosciences,
DOI 10.1007/978-3-642-18015-6_12, © Springer-Verlag Berlin Heidelberg 2011

that they produce without will (Scepkowski and Cronin-Golomb 2003); others can lose the subjective experience of wanting to move (Sirigu et al. 2004). With respect to awareness, some subjects obstinately claim that they are moving a paralyzed arm (Orfei et al. 2007). Others report movements in a limb that no longer exists (Ramachandran and Hirstein 1998).

The identification of intention and awareness as valid objects for scientific exploration triggered a large number of studies in healthy subjects and brain-damaged patients (Desmurget and Sirigu 2009; Haggard 2005, 2008). Based on these studies, two main hypotheses could be put forward: (1) the motor system is not aware of what it is doing but rather of what it intended to do; and (2) intention and awareness share a common neural substrate within the posterior parietal cortex (PPC). These two hypotheses are discussed in the present review.

2 Motor Intention Shapes Movement Awareness

The motor system operates mainly at an unconscious level, as can be seen in postural regulations (Wing et al. 1997), non-verbal communication skills (Dijksterhuis and Bargh 2001), visual scanning of complex scenes (Yarbus 1967), on-line control of reaching movements (Desmurget and Grafton 2003), and behavioral studies showing that the motor system can remain unaware of large sensory distortions occurring during task realization. A good example of this fact was provided recently by Turner and colleagues (2003), who required human subjects to perform a tracking task. During task performance, a white circle moved horizontally across a monitor at a constant speed (10 cm/s) between endpoints 20 cm apart. The circle reversed direction of movement with no delay on reaching left and right endpoints. Subjects were instructed to keep the circle within a red square controlled by a hand-held joystick. On different sessions, the gain of the relationship between joystick movement and cursor displacement was modified in such a way that joystick displacements of 6, 12, 18, and 24 cm produced cursor displacements of 20 cm. None of the subjects involved in the study exhibited awareness that the joystick-to-cursor relationship had changed from session to session.

A comparable observation had been reported a few years earlier by Fourneret and Jeannerod (1998). These authors required healthy humans to trace sagittal lines on a graphic tablet. Visual feedback of the movement was provided to the subjects through a mirror positioned above the tablet. In some trials, this feedback was shifted so that the line traced by the subjects deviated to the right or the left by a substantial amount (up to 10°). To perform a straight movement, the subjects had thus to produce a lateral response. They were able to do so quite easily. However, they kept reporting that their movement was straight in the sagittal direction. They remained unaware of their motor adjustment, suggesting, in the authors' words, that "normal subjects are not aware of signals generated by their own movements" (p. 1133). Clear evidence supporting this conclusion is provided by the so-called subliminal double-step task. In this paradigm, the subjects are required to "look and point" to visual targets displayed in the peripheral visual field. During saccadic gaze displacement, when conscious visual perception is abolished, the target location is modified. This intra-saccadic jump is quickly taken into account to modify the hand-reaching response. Strikingly, the kinematic modifications applied to the ongoing response are not consciously detected by the subjects, who remain completely unaware of the profound changes that affect path curvature and individual joint trajectories (Desmurget and Grafton 2000; Desmurget et al. 2001, 2004).

Based on these data, it is tempting to speculate that the motor system is not aware of its ongoing production but rather of its initial intention. As long as the goal is achieved, nothing reaches awareness about the details of the ongoing movements, even when substantial corrections have to be implemented to attain the intended state. If so, one may predict that motor awareness is mainly mediated by efferent signals. Three main lines of evidence support this prediction. The first one is associated with studies aimed at investigating the issue of self recognition. In one of these studies, passive movements of the right index finger were investigated (Tsakiris et al. 2005). This finger was moved through a lever operated by the subject or by the left hand of the experimenter. Visual feedback about the movement was manipulated so that subjects observed their own or someone else's right hand. Self recognition was significantly more accurate when the subjects were the authors of the action, i.e., when an efferent output was generated, showing that efferent information is more important than sensory signals for constructing motor awareness in the context of self-generated actions.

The second line of evidence suggesting that efferent signals are important for the emergence of motor awareness comes from studies in which the subjects reported being aware of performing a movement, although no sensory signal was present. Recently, Kristeva and colleagues (2006) required a deafferented patient (GL) to perform self-paced flexion of the index finger. In control subjects, this task triggered contralateral movement-evoked potentials in the sensorimotor area. No such response was found in GL. However, this absence of sensory input did not prevent the patient from being aware of her movement. She knew that she was moving, which indicated, in the authors' words, that "she had a normal motor awareness" (p. 684). Of course, she had no "perceptual awareness" in the sense that she had no "feeling" about her movement. In fact, GL was "aware" that she was moving but she could not determine whether she was moving as expected. A similar conclusion was reached by Lafargue and colleagues (2003) in another study. These authors required GL to produce a target force with her right hand and then match this force with her left hand. Despite variations in the motor command, GL was able to perform the task with good accuracy. She was aware not only that she was moving but also of the level of force that she was applying.

The third (and probably the most compelling) line of evidence suggesting that motor awareness has a central origin comes from the case of hemiplegic patients with anosognosia (Bisiach and Geniniani 1991; Orfei et al. 2007). These patients claim stubbornly that they are moving despite their paralysis. A representative example is provided by Ramachandran in a well-known review (1996):

Doctor: “Can you clap?” Patient: “Of course I can clap.” Doctor: “Will you clap for me?” At this point, the patient initiates clapping movements with the right hand, as if clapping with an imaginary hand near the sagittal plane. The discussion resumes. Doctor: “Are you clapping?” Patient: “Yes, I am clapping.” To explain this result, it is often suggested that the brain mechanisms that normally compare the expected and actual peripheral reafferences are damaged in hemiplegic patients with anosognosia, which prevents the subjects from knowing that they are not moving (Berti et al. 2005). This idea was recently tested in a study carried out by Fotopoulou and his colleagues (2008). Paralyzed patients with and without anosognosia for hemiplegia were given false visual feedback of movement in their left, paralyzed arm through a prosthetic rubber hand. Three conditions were investigated: (1) the patients were instructed to raise their paralyzed arm, (2) the patients were told that the experimenter would move their arm; and (3) no instruction. In contrast to hemiplegic patients without anosognosia, paralyzed patients with anosognosia disregarded visual information of their motionless rubber hand when they had the intention to move compared to when they expected the experimenter to move the rubber hand, or had no movement expectation at all. These findings clearly support the idea that motor awareness derives from the processing of motor intentions.

Of course, the data above do not mean that sensory signals play no role in motor awareness. They just mean that this role is limited to specific situations where our motor predictions prove dramatically unreliable, which is rarely the case at least in adults. Indeed, we rarely see our hand going rightward or downward when we move it leftward or upward. Similarly, we seldom, if ever, face the surprise of seeing our arm not responding to a motor command. When discrepancies of this sort occur, the motor control system can no longer force the congruence between the desired and actual sensory reafferences, as it does continuously for errors of limited magnitudes. In this case, an error signal is emitted and we become aware of our movements as they actually unfold. In paralyzed patients with anosognosia, this ability to process the sensory signals and compare them to the efferent output appears to be abolished, due to a lesion of the premotor cortex (Desmurget and Sirigu 2009; Berti et al. 2005).

3 Motor Intention and Movement Awareness Share a Common Parietal Substrate

During the last decade, solid evidence has linked the feeling of conscious intention to the activity of the PPC. For instance, it was reported that selective lesions of this region can cause alien hand movements(Assal et al. 2007). Also, it was shown that patients with parietal lesions could lose the early subjective experience of wanting to move. This finding was demonstrated by Sirigu and colleagues (2004) in a behavioral experiment using the so-called Libet’s paradigm. In this paradigm, subjects have to fixate a spot rotating on a screen. They initiate a voluntary press-button movement with their right index finger whenever they feel a desire to do so. At a random time after this movement, the rotating spot is stopped, indicating that the subjects should report where the spot was when they first felt their desire to move ("Will to move", W-Judgment). In their original paper, Libet and his colleagues (1983) found that the W-Judgment occurred 206 ms before EMG onset in normal subjects. A slightly longer delay (239 ms) was reported by Sirigu and colleagues. Interestingly, this delay was almost abolished in patients with parietal lesions, suggesting that the patients did not know about their intention to move until movement release became imminent.

Recently, these data were confirmed in a study wherein direct electrical stimulation was applied over the parietal cortex in patients undergoing awake surgery for tumor removal (Desmurget et al. 2009). The authors of this study reasoned that, if the PPC is truly involved in the generation of conscious intentions, then electrical stimulation of this region should be accompanied by the subjective experience of "wanting to move". In agreement with this prediction, stimulation of the right inferior parietal lobule [Brodmann Areas (BA) 40 and 39] was found to trigger a strong desire to move the contralesional hand, arm or foot. In the same vein, stimulations of the left inferior parietal lobule (BA 39) were reported to provoke an intention to move the lips. Often, the patients reported "a will to move the chest" or "a desire to move the hand", which demonstrated the voluntary and endogenous character of the feelings they experienced. A typical exchange was as follows. Experimenter: "Did you move?" Patient: "No... I had a desire to roll my tongue in my mouth." Experimenter: "To roll what, your...?" Patient: "To roll my tongue in my mouth." To account for these observations, it was suggested that the PPC contains stored movement representations (Sirigu et al. 1995; Rothi et al. 1997) and that electrical stimulations activate these representations, thus provoking a desire to move (Desmurget and Sirigu 2009; Desmurget et al. 2009).

Interestingly, in the stimulation study described above, conscious intention gave rise to illusory movements when the intensity of the stimulation was raised. In this situation, the subjects become aware of movements that did not occur. They reported that they had moved their arm, mouth or foot in the absence of EMG signal and overt movement. For instance, the verbal exchange reported above progressed as follows when the intensity of the stimulation was raised from 5 to 8 mA. Experimenter: "Did you move?" Patient: "Yes, yes, a corner of the mouth." Experimenter: "You did move the mouth?" Patient: "Yes." To account for these observations, it was suggested that the PPC contains stored movement representations (Sirigu et al. 1995; Rothi et al. 1997) and that electrical stimulations activate these representations, thus provoking a desire to move (Desmurget and Sirigu 2009; Desmurget et al. 2009). To account for these findings, it was suggested that higher intensities of stimulation did not just prime a motor representation to consciousness (giving rise to a conscious intention) but also recruited the executive network responsible for movement monitoring. This network is assumed to rely on forward modeling, a process that simulates the effect of the efferent output to estimate, in real time, the state of the motor apparatus (Wolpert and Flanagan

2001; Desmurget and Grafton 2000, 2003). There is converging evidence that forward models are generated in the posterior parietal lobe (Sirigu et al. 1999; Wolpert et al. 1998; Pellijeff et al. 2006; Ogawa et al. 2007).

Beyond these data, several studies have confirmed the role of the PPC in motor awareness. For instance, it was reported that a patient presenting with damage to the posterior parietal region on the left side became unable to maintain a conscious representation of her right limbs over time. This patient reported feelings like "losing her right arm". Also, when she was in a bus, she was sometimes surprised to "find" her right leg in the middle of the aisle, as other passengers tripped over her foot (Wolpert et al. 1998). A comparable observation was reported by Sirigu and colleagues in a behavioral study (Sirigu et al. 1999) in which the patients were required to perform various hand movements. These movements were recorded with a video camera and fed back to the patients through a mirror positioned above their hand. However, in some trials, the hand displayed in the mirror was not the real hand of the patients but the hand of the experimenter executing a similar response. Results indicated that the patients were more impaired than healthy subjects at recognizing their own hand. Of particular interest were the trials in which the patients produced highly inaccurate and clumsy gestures. In nearly 90% of these trials, the patients believed that they were observing their own hand when watching a smooth and accurate movement performed by the experimenter. This finding could be explained by the fact that the movements executed by the experimenter closely matched the goal of the movement the patients were originally instructed to perform.

4 Conclusions

In this paper, we have briefly reviewed evidence that the motor system is mainly aware of its intention. As long as we reach our goal, nothing enters awareness regarding the kinematic details of the ongoing movements, even when substantial corrections have to be implemented. This phenomenological link between motor intention and movement awareness is also present at the neural level, where intention and awareness share a common anatomical substrate in the PPC.

References

Assal F, Schwartz S, Vuilleumier P (2007) Moving with or without will: functional neural correlates of alien hand syndrome. Ann Neurol 62:301–306

Berti A, Bottini G, Gandola M, Pia L, Smania N, Stracciari A, Castiglioni I, Vallar G, Paulesu E (2005) Shared cortical anatomy for motor awareness and motor control. Science 309:488–491

Bisiach E, Geniniani G (1991) Anosognosia related to hemiplegia and heminopia. In: Prigatano GP, Scahcter DL (eds) Awareness of deficit after brain injury. Oxford University Press, New York, pp 17–39

Desmurget M, Grafton S (2000) Forward modeling allows feedback control for fast reaching movements. Trends Cogn Sci 4:423–431

Desmurget M, Grafton S (2003) Feedback or forward control: end of a dichotomie. In: Johnson S (ed) Cognitive neuroscience perspectives on the problem of intentional action. MIT, Boston, MA, pp 289–338

Desmurget M, Sirigu A (2009) A parietal-premotor network for movement intention and motor awareness. Trends Cogn Sci 13:411–419

Desmurget M, Grea H, Grethe JS, Prablanc C, Alexander GE, Grafton ST (2001) Functional anatomy of nonvisual feedback loops during reaching: a positron emission tomography study. J Neurosci 21:2919–2928

Desmurget M, Gaveau V, Vindras P, Turner RS, Broussolle E, Thobois S (2004) On-line motor control in patients with Parkinson's disease. Brain 127:1755–1773

Desmurget M, Reilly KT, Richard N, Szathmari A, Mottolese C, Sirigu A (2009) Movement intention after parietal cortex stimulation in humans. Science 324:811–813

Dijksterhuis A, Bargh JA (2001) The perception-behavior expressway: automatic effects of social perception on social behavior. Adv Exp Soc Psychol 33:1–40

Fotopoulou A, Tsakiris M, Haggard P, Vagopoulou A, Rudd A, Kopelman M (2008) The role of motor intention in motor awareness: an experimental study on anosognosia for hemiplegia. Brain 131:3432–3442

Fourneret P, Jeannerod M (1998) Limited conscious monitoring of motor performance in normal subjects. Neuropsychologia 36:1133–1140

Frith CD, Blakemore SJ, Wolpert DM (2000) Abnormalities in the awareness and control of action. Philos Trans R Soc Lond B Biol Sci 355:1771–1788

Haggard P (2005) Conscious intention and motor cognition. Trends Cogn Sci 9:290–295

Haggard P (2008) Human volition: towards a neuroscience of will. Nat Rev Neurosci 9:934–946

Kristeva R, Chakarov V, Wagner M, Schulte-Monting J, Hepp-Reymond MC (2006) Is the movement-evoked potential mandatory for movement execution? A high-resolution EEG study in a deafferented patient. Neuroimage 31:677–685

Lafargue G, Paillard J, Lamarre Y, Sirigu A (2003) Production and perception of grip force without proprioception: is there a sense of effort in deafferented subjects? Eur J Neurosci 17:2741–2749

Libet B, Gleason CA, Wright EW, Pearl DK (1983) Time of conscious intention to act in relation to onset of cerebral activity (readiness-potential). The unconscious initiation of a freely voluntary act. Brain 106(Pt 3):623–642

Ogawa K, Inui T, Sugio T (2007) Neural correlates of state estimation in visually guided movements: an event-related fMRI study. Cortex 43:289–300

Orfei MD, Robinson RG, Prigatano GP, Starkstein S, Rüsch N, Bria P, Caltagirone C, Spalletta G (2007) Anosognosia for hemiplegia after stroke is a multifaceted phenomenon: a systematic review of the literature. Brain 130:3075–3090

Pellijeff A, Bonilha L, Morgan PS, McKenzie K, Jackson SR (2006) Parietal updating of limb posture: an event-related fMRI study. Neuropsychologia 44:2685–2690

Ramachandran VS (1996) What neurological syndromes can tell us about human nature: some lessons from phantom limbs, capgras syndrome, and anosognosia. Cold Spring Harb Symp Quant Biol 61:115–134

Ramachandran VS, Hirstein W (1998) The perception of phantom limbs. The D.O. Hebb lecture. Brain 121(Pt 9):1603–1630

Rothi LJ, Ochipa C, Heilman KM (1997) A cognitive neurpsychological model of limb praxis and apraxia. In: Rothi LJ, Heilman KM (eds) Apraxia: the neuropsychology of action. Psychology Press, Hove, pp 29–49

Scepkowski LA, Cronin-Golomb A (2003) The alien hand: cases, categorizations, and anatomical correlates. Behav Cogn Neurosci Rev 2:261–277

Sirigu A, Cohen L, Duhamel JR, Pillon B, Dubois B, Agid Y (1995) A selective impairment of hand posture for object utilization in apraxia. Cortex 31:41–55

Sirigu A, Daprati E, Pradat-Diehl P, Franck N, Jeannerod M (1999) Perception of self-generated movement following left parietal lesion. Brain 122(Pt 10):1867–1874
Sirigu A, Daprati E, Ciancia S, Giraux P, Nighoghossian N, Posada A, Haggard P (2004) Altered awareness of voluntary action after damage to the parietal cortex. Nat Neurosci 7:80–84
Tsakiris M, Haggard P, Franck N, Mainy N, Sirigu A (2005) A specific role for efferent information in self-recognition. Cognition 96:215–231
Turner RS, Desmurget M, Grethe J, Crutcher MD, Grafton ST (2003) Motor subcircuits mediating the control of movement extent and speed. J Neurophysiol 90:3958–3966
Wing AM, Flanagan JR, Richardson J (1997) Anticipatory postural adjustments in stance and grip. Exp Brain Res 116:122–130
Wolpert DM, Flanagan JR (2001) Motor prediction. Curr Biol 11:R729–R732
Wolpert DM, Goodbody SJ, Husain M (1998) Maintaining internal representations: the role of the human superior parietal lobe. Nat Neurosci 1:529–533
Yarbus AL (1967) Eye movements and vision. Plenum Press, New York

Index

S. Dehaene and Y. Christen (eds.), *Characterizing Consciousness: From Cognition to the Clinic?* Research and Perspectives in Neurosciences,
DOI 10.1007/978-3-642-18015-6,

Printing: Ten Brink, Meppel, The Netherlands
Binding: Stürtz, Würzburg, Germany